Lie Groups and Lie Algebras:
A Physicist's Perspective

Lie Groups and Lie Algebras: A Physicist's Perspective

Adam M. Bincer
Professor Emeritus
University of Wisconsin–Madison

OXFORD
UNIVERSITY PRESS

Great Clarendon Street, Oxford, OX2 6DP,
United Kingdom

Oxford University Press is a department of the University of Oxford.
It furthers the University's objective of excellence in research, scholarship,
and education by publishing worldwide. Oxford is a registered trade mark of
Oxford University Press in the UK and in certain other countries

© Adam M. Bincer 2013

The moral rights of the author have been asserted

First Edition published in 2013

Impression: 1

All rights reserved. No part of this publication may be reproduced, stored in
a retrieval system, or transmitted, in any form or by any means, without the
prior permission in writing of Oxford University Press, or as expressly permitted
by law, by licence or under terms agreed with the appropriate reprographics
rights organization. Enquiries concerning reproduction outside the scope of the
above should be sent to the Rights Department, Oxford University Press, at the
address above

You must not circulate this work in any other form
and you must impose this same condition on any acquirer

British Library Cataloguing in Publication Data

Data available

Library of Congress Cataloging in Publication Data

Library of Congress Control Number: 2012943007

ISBN 978–0–19–966292–0

Printed and bound by
CPI Group (UK) Ltd, Croydon, CR0 4YY

In memory of Wanda

Preface

This book grew out of lecture notes for a course on group theory that I taught to graduate students in Physics at the University of Wisconsin–Madison. I myself learned much of this material from Charlie Goebel by sitting in on this class when he taught it before me. I want to take this opportunity to thank Charlie for his generosity in answering my endless questions as I was writing this book and to express my admiration for his command of the subject. Since group theory has been around for quite some time there is really nothing new in this book other than my approach to the subject—an example might be my treatment of the Wigner–Eckart theorem and the reduced matrix element in Chapter 4.

Hermann Weyl is quoted by Freeman Dyson to have said "My work always tried to unite the true with the beautiful; but when I had to choose one or the other, I usually chose the beautiful". This is, perhaps, going too far but I also believe that mathematics, and particularly group theory, is beautiful. To a large extent my justification for writing this book is to convey that feeling. To see what I mean consider $SO(2)$, the orthogonal group in two dimensions. This group is Abelian and therefore all its irreducible representations are one-dimensional. Yet the defining (vector) representation is obviously two-dimensional. The resolution of this dilemma by the duality concept—for details see Chapter 7—is an example of this beauty at work.

At the end of every chapter in which a mathematician or physicist is mentioned for the first time I have included a short biographical sketch of that person. My hope is that the reader will be intrigued and pursue a more detailed biography of these individuals on whose shoulders we stand today.

After providing in Chapters 1 and 2 the definitions of many of the concepts that will be needed I deal in Chapters 3 and 4 with rotations. This is material that most physicists learn when they deal with angular momentum and I use here group theory language to discuss matters familiar in a different setting.

The preceding leads naturally to the discussion of rotations in n dimensions, which is the content of Chapters 5 through 8. Whereas the tensor representations of $SO(n)$ are a natural generalization of the orbital angular momentum, the $Spin(n)$ groups (i.e. the universal covering groups of the $SO(n)$ groups) also have spinor representations corresponding to generalizing spin angular momentum. These spinor representations are introduced using Clifford numbers. Although in general only their abstract properties are needed we provide an explicit construction in terms of $2^m \times 2^m$ matrices, which simplifies the demonstration of which spinor representations are complex, orthogonal or symplectic.

The Clifford numbers lead naturally to the discussion of composition algebras, Hurwitz's theorem, quaternions and octonions (Chapter 9), which in turn leads to the discussion of the exceptional group G_2 as the group of automorphisms of octonions (Chapter 10).

In Chapter 11 I return to the orthogonal groups to discuss their Casimir operators in a convenient matrix notation. Here I make the important point that for the orthogonal groups in even dimensions a Casimir operator constructed in a different way must be included in the integrity basis.

In Chapter 12 unitary and symplectic groups are defined together with the orthogonal groups as classical groups and their dimensions and connectivity are obtained in one fell swoop by viewing them as unitary groups over the field of $\mathbb{R}$, $\mathbb{C}$ and $\mathbb{H}$, i.e. the real, the complex and the quaternion. The unitary groups are discussed further in Chapters 13, 14 and 15 where, in particular, in discussing irreducible representations of $SU(n)$ the symmetric group is introduced and dealt with using the Young Tableaux.

Chapters 16 and 17 describe the Cartan basis and the Cartan classification of semisimple algebras while Dynkin diagrams are introduced in Chapter 18.

Space-time symmetries are discussed in Chapter 19, where I describe representations of the Lorentz group, and Chapter 20, where I describe representations of the Poincaré group. At the end of that chapter I describe conformal invariance and briefly touch on Virasoro and Kac–Moody algebras.

In the last chapter I generalize to n space dimensions Pauli's group-theoretical treatment of the energy levels of the non-relativistic hydrogen atom.

There are many books on group theory, among those that look at group theory from the point of view of physicists I would like to mention *Lie Groups, Lie Algebras and Some of Their Applications* (1974) by

R. Gilmore, *Group Theory in Physics* (1985) by W. Tung and *Group Theory, a Physicist's Survey* (2010) by P. Ramond. The choice of topics in my book is somewhat more modern than Gilmore's and Tung's, for example I consider octonions, which they do not. Ramond's book is more sophisticated—again using octonions for comparison, whereas I only connect them to G_2 he discusses their relation to all of the exceptional groups.

Contents

Chapter 1. Generalities   1
Definitions of group, isomorphism, representation, vector space and algebra. Biographical notes on Galois, Abel and Jacobi are given.

Chapter 2. Lie groups and Lie algebras   8
Lie Groups, infinitesimal generators, structure constants, Cartan's metric tensor, simple and semisimple groups and algebras, compact and non-compact groups. Biographical notes on Euler, Lie and Cartan are given.

Chapter 3. Rotations: $SO(3)$ and $SU(2)$ 18
Rotations and reflections, connectivity, center, universal covering group.

Chapter 4. Representations of $SU(2)$ 26
Irreducible representations, Casimir operators, addition of angular momenta, Clebsch–Gordan coefficients, the Wigner–Eckart theorem, multiplicity. Biographical notes on Casimir, Weyl, Clebsch, Gordan and Wigner are given.

Chapter 5. The $so(n)$ algebra and Clifford numbers   42
$Spin(n)$, spinors and semispinors, Schur's lemma. Biographical notes on Clifford and Schur are given.

Chapter 6. Reality properties of spinors   53
Conjugate, orthogonal and symplectic representations.

Chapter 7. Clebsch–Gordan series for spinors   58
Antisymmetric tensors, duality.

Chapter 8. The center and outer automorphisms of $Spin(n)$ 68
Inversion, $\mathbb{Z}_2$, $\mathbb{Z}_4$ and $\mathbb{Z}_2 \times \mathbb{Z}_2$ centers. A biographical note on Dynkin is given.

Chapter 9. Composition algebras   74
Hurwitz's theorem, quaternions and octonions, non-associativity. Biographical notes on Hurwitz, Hamilton, Graves, Cayley and Frobenius are given.

Chapter 10. The exceptional group G_2 84
Automorphisms of octonions, quaternions and complex numbers. A biographical note on Racah is given.

Chapter 11. Casimir operators for orthogonal groups   89
The invariant ε tensor, integrity basis. A biographical note on Pfaff is given.

Chapter 12. Classical groups   93
Orthogonal, symplectic and unitary groups; their dimensions and connectivity. Biographical notes on Lorentz, de Sitter, Liouville, Maxwell and Thomas are given.

Chapter 13. Unitary groups   102
Generators, tensors, Casimir operators; anomaly cancellation.

Chapter 14. The symmetric group S_r and Young tableaux   107
Permutations, the symmetric group, its representations and their dimension, Young patterns and tableaux. A biographical note on Young is given.

Chapter 15. Reduction of $SU(n)$ tensors   117
Irreducible tensors and their dimensions, conjugate representations, reduction of the Kronecker product. A biographical note on Kronecker is given.

Chapter 16. Cartan basis, simple roots and fundamental weights   129
Cartan subalgebra, Weyl reflections.

Chapter 17. Cartan classification of semisimple algebras   139
The rank one A_1 algebra, the rank two $A_2, B_2 = C_2, D_2 = A_1 \oplus A_1$ and G_2 algebras, the rank n A_n, B_n, C_n and D_n series of algebras, the F_4 and the E_n, $6 \leq n \leq 8$, exceptional algebras.

Chapter 18. Dynkin diagrams 151
Rules for constructing Dynkin diagrams, simply laced algebras, folding.

Chapter 19. The Lorentz group 162
Lorentz Casimirs, principal and complementary series of unitary representations, finite-dimensional representations, the universal cover $SL(2, \mathbb{C})$, polar decomposition. Biographical notes on Minkowski, Klein, Gordon, Dirac and Proca are given.

Chapter 20. The Poincaré and Liouville groups 175
Semidirect product, Pauli–Lubanski four-vector, Poincaré Casimirs, representations of the Poincaré group, conformal group, Virasoro and Kac–Moody algebras. Biographical notes on Poincaré, Pauli, Lubanski, Kac and Moody are given.

Chapter 21. The Coulomb problem in n space dimensions 189
Heisenberg algebra, Lenz–Runge n-vector, energy levels of the hydrogen atom. Biographical notes on Coulomb, Heisenberg, Lenz and Runge are given.

Bibliography and References 196

Index 199

1
Generalities

We suppose it is reasonable to start with the definition of the word **group**, introduced into mathematics in its technical sense in 1830 by Galois. A group consists of a set of **elements** ($a,b,c, \ldots$) and a **composition law**, which tells us how two elements are combined to form another element. We shall refer to the composition law as **multiplication** but it must be understood that this term is symbolic.

The composition law must be **associative**:

$$a(bc)=(ab)c=abc, \tag{1.1}$$

there must exist an element e, called the **identity**, (which, as is easy to show, is unique) with the properties

$$ae=ea=a, \tag{1.2}$$

every element a must have an **inverse** a^{-1} such that

$$a^{-1}a = a\ a^{-1} = e \tag{1.3}$$

and we must have **closure**: if a and b are elements of the group then so is c, where $c=ab$.

If the composition law is **commutative**, meaning

$$ab=ba \tag{1.4}$$

for every a and b in the group then the group is called **Abelian**; otherwise, if for some a and b, $ab \neq ba$, it is called **non-Abelian**.

A **subgroup** H of a group G is a group whose elements h are a subset of the elements g in G and a subgroup is called **proper** if it is not simply the identity element or the entire group. A subgroup H of a group G is called an **invariant** subgroup if for every $h \in H$ and every $g \in G$ we have

$$ghg^{-1} \in H. \tag{1.5}$$

An example of a group is the infinite *additive group of integers*. Here the elements are all integers: ...−3, −2, −1, 0, 1, 2,..., and the composition law is ordinary addition. Clearly we have

$$a+(b+c)=(a+b)+c=a+b+c, \qquad (1.6)$$

$$a+0=0+a=a, \qquad (1.7)$$

$$a-a=-a+a=0, \qquad (1.8)$$

$$a+b=b+a \Rightarrow \text{Abelian}. \qquad (1.9)$$

The cyclic group $\mathbb{Z}_n$ provides another example. This is a finite group of **order** n, meaning that it has n distinct elements $a_k, k=0,1,...,n-1$, with a_0 the identity, and $a_k = a_1 a_1 ... a_1 = (a_1)^k$ and $a_n = (a_1)^n = a_0$. One says that this group is generated by a_1 (the word **cyclic** in this definition refers to the fact that this group is generated by a single element). This group is obviously Abelian and a concrete example of it is the *addition of integers modulo n*: $a_k \to k \mod n$. On the other hand, consider the nth roots of unity:

$$\exp\frac{2ik\pi}{n}, \quad k=0, 1,..., n-1, \qquad (1.10)$$

with the composition law being ordinary multiplication—this is precisely $\mathbb{Z}_n$ again.

We have here an explicit example of an interesting feature of groups: our definition of $\mathbb{Z}_n$ was merely a statement of its multiplication table and that is all. The addition of integers modulo n or the multiplication of nth roots of unity are realizations of the cyclic group $\mathbb{Z}_n$—these two realizations are completely different *concretely*, yet of course they are *abstractly* the same thing. When we have two such different realizations we say that the two groups are isomorphic. A mapping of one algebraic structure into another is called a **homomorphism** if it preserves all combinatorial operations associated with that structure. If the mapping is in addition one-to-one, or faithful, so that it could have been applied just as well in reverse, then it is called an **isomorphism**.

Note that the abstract multiplication table is

$$a_p \, a_s = a_{p+s} \qquad (1.11)$$

(all subscripts taken mod n), which is realized in the first case by

$$p_{\text{mod } n} + s_{\text{mod } n} = (p+s)_{\text{mod } n} \tag{1.12}$$

and in the second case by

$$\exp\frac{2ip\pi}{n}\exp\frac{2is\pi}{n} = \exp\frac{2i(p+s)\pi}{n}, \tag{1.13}$$

i.e. if $D(a)$ corresponds to element a then we must have

$$D(a)D(b) = D(ab). \tag{1.14}$$

Our examples above were Abelian. As is well known, matrix multiplication is non-commutative and so not surprisingly expressing group elements by matrices is often an excellent way to obtain concrete descriptions of non-Abelian groups. When a group is described in some concrete manner we say that we have a **realization**, and if this realization is in terms of matrices we say that we have a **representation**.

If the number of group elements is finite we have a **finite discrete** group. If the elements are denumerably infinite then we have an **infinite discrete** group. If the elements form a continuum we have a **continuous** group.

Now that we have defined the concept of a group we consider a slightly more complicated object, a field. Whereas a group is a set of elements with one binary operation defined on them, by a **field** F we mean a collection of elements closed under *two* binary operations, conventionally called addition and multiplication, such that

i) under addition, F is an Abelian group with the identity element denoted by 0;
ii) under multiplication, F with 0 omitted is a group with identity element denoted by 1;
iii) addition is distributive under multiplication, i.e.

$$a(b+c) = ab + ac. \tag{1.15}$$

Familiar examples of fields are the *rationals*, the *reals*, and the *complex* numbers. For each of these fields multiplication is commutative. The *quaternions* form a field with non-commutative multiplication. The *integers* are *not* a field because the multiplicative inverse of an integer is not an integer.

We next define a **linear vector space** V over a field F as a collection of elements $\mathbf{v}_1, \mathbf{v}_2,\ldots \in V$ called vectors *and* a field F, whose elements f are called scalars, and two kinds of operations called **vector addition** and **scalar multiplication**, such that

i) the vectors form an Abelian group under vector addition;
ii) for $\mathbf{v}\in V$, $f\in F$ we have scalar multiplication obeying

$$f\mathbf{v}\in V,$$
$$f_1(f_2\mathbf{v})=(f_1f_2)\mathbf{v},$$
$$1\mathbf{v}=\mathbf{v}1=\mathbf{v},$$
$$f(\mathbf{v}_1+\mathbf{v}_2)=f\mathbf{v}_1+f\mathbf{v}_2,$$
$$(f_1+f_2)\mathbf{v}=f_1\mathbf{v}+f_2\mathbf{v}. \tag{1.16}$$

An important concept here is that of a **basis**, which means a complete set of linearly independent vectors. The vectors $\mathbf{v}_1, \mathbf{v}_2,\ldots, \mathbf{v}_n$ are **linearly independent** if

$$\Sigma_{1\leq i\leq n}f_i\mathbf{v}_i=0 \Rightarrow f_i=0 \tag{1.17}$$

and they form a **complete** set if any vector can be expressed in terms of them. A vector space is called N-dimensional if it is possible to find in it N linearly independent vectors but not $N+1$; any such maximal set of N vectors provides a basis.

The $n\times n$ real matrices are an example of a vector space. This is the space whose "vectors" are the matrices and vector addition is ordinary matrix addition. The "scalars" are the reals and scalar multiplication is ordinary multiplication of matrices by numbers.

This space is n^2-dimensional and a basis is provided by the n^2 matrices with zeros everywhere except for a 1 in a single place, a different place for each of the n^2 matrices. For example, for $n=2$, a basis is the following four matrices:

$$\begin{pmatrix} 1 & 0 \\ 0 & 0 \end{pmatrix} \begin{pmatrix} 0 & 1 \\ 0 & 0 \end{pmatrix} \begin{pmatrix} 0 & 0 \\ 1 & 0 \end{pmatrix} \begin{pmatrix} 0 & 0 \\ 0 & 1 \end{pmatrix}. \tag{1.18}$$

A *canonical* realization of an N-dimensional vector space is in terms of $N\times 1$ matrices with basis vectors

$$\mathbf{e}_1 = \begin{pmatrix} 1 \\ 0 \\ 0 \\ . \\ . \\ 0 \\ 0 \end{pmatrix}, \quad \mathbf{e}_2 = \begin{pmatrix} 0 \\ 1 \\ 0 \\ . \\ . \\ 0 \\ 0 \end{pmatrix}, \ldots, \quad \mathbf{e}_N = \begin{pmatrix} 0 \\ 0 \\ 0 \\ . \\ . \\ 0 \\ 1 \end{pmatrix} \qquad (1.19)$$

and all N-dimensional vector spaces over the same field F are isomorphic.

A **linear algebra** A is a vector space V in which an additional operation called **vector multiplication** (which we denote by #) is defined such that

$$\mathbf{u},\mathbf{v},\mathbf{w} \in V \Rightarrow \mathbf{u}\#\mathbf{v} \in V,$$
$$(\mathbf{u}+\mathbf{v})\#\mathbf{w} = \mathbf{u}\#\mathbf{w}+\mathbf{v}\#\mathbf{w}, \qquad (1.20)$$
$$\mathbf{u}\#(\mathbf{v}+\mathbf{w}) = \mathbf{u}\#\mathbf{v}+\mathbf{u}\#\mathbf{w}.$$

Table 1.1 indicates the increasing complexity of concepts:

Table 1.1 Concepts

	Elements	*Operations*
set	one kind	none
group	one kind	one kind: multiplication
field	one kind	two kinds: addition and multiplication
vector space	two kinds: vectors and scalars	two kinds: vector addition and scalar multiplication
algebra	two kinds: vectors and scalars	three kinds: vector addition, scalar multiplication and vector multiplication

Now an algebra can have additional properties and then it is given special names. The algebra is called **associative** if $(\mathbf{u}\#\mathbf{v})\#\mathbf{w}=\mathbf{u}\#(\mathbf{v}\#\mathbf{w})$. It is said to possess an **identity 1** (in general different from the **0** or the 1, the identities under vector addition and scalar multiplication) if

v#1=1#v=v. It is said to be **symmetric** if **v#w=w#v**, **antisymmetric** if **v#w=−w#v**, and finally it is said to have the **derivative** property if **u#(v#w)=(u#v)#w+v#(u#w)**.

For our first example we consider $n \times n$ real matrices, which form a real n^2-dimensional vector space under matrix addition and multiplication by reals. If we define # to be *matrix multiplication* we obtain an algebra that is associative and has the identity $(\mathbf{1})_{ik}=\delta_{ik}$. [Note that this identity is different from $(\mathbf{0})_{ik}=0$, the identity under matrix addition, or from 1, the identity under scalar multiplication.]

For our second example we consider the set of real $n \times n$ symmetric matrices. A matrix S is called **symmetric** if

$$S^{\mathrm{T}}=S, \tag{1.21}$$

where the superscript T denotes transpose, which is the operation that flips a matrix across its main diagonal, i.e. for any matrix M we have

$$(M^{\mathrm{T}})_{ik}=(M)_{ki}. \tag{1.22}$$

Clearly this set is a subset of the vector space of our first example and is itself a vector space. If we now define # to be matrix multiplication we do *not* get an algebra because the matrix obtained by multiplying two symmetric matrices is not itself symmetric in general:

$$(S_1 S_2)^{\mathrm{T}}=S_2{}^{\mathrm{T}} S_1{}^{\mathrm{T}}=S_2 S_1 \neq S_1 S_2. \tag{1.23}$$

If, however, we define # to be symmetrized matrix multiplication:

$$S_1 \# S_2 = S_1 S_2 + S_2 S_1 \equiv \{S_1, S_2\} \tag{1.24}$$

then we do get an algebra—one readily verifies that it is not associative, and not derivative, but it does have an identity, namely $\frac{1}{2}\mathbf{1}$. The structure $\{S_1, S_2\}$, defined above, is called the **anticommutator** of S_1 and S_2.

As a third example we consider real $n \times n$ antisymmetric matrices A satisfying

$$A^{\mathrm{T}}=-A, \tag{1.25}$$

which is again a subset of all $n \times n$ real matrices, which is again *not* closed under matrix multiplication. If, however, we define # to be antisymmetrized matrix multiplication:

$$A_1 \# A_2 = A_1 A_2 - A_2 A_1 \equiv [A_1, A_2] \tag{1.26}$$

then this system is an algebra, which is not associative, does not have an identity, but does have the derivative property

$$[A_1,[A_2,A_3]]=[[A_1,A_2],A_3]+[A_2,[A_1,A_3]], \qquad (1.27)$$

which can be verified by explicit calculation (we assume here that a product like $A_1 A_2$ is defined by ordinary matrix multiplication). The structure $[A_1,A_2]$, defined above, is called the **commutator** of A_1 and A_2 and (1.27) is the **Jacobi identity** applied to commutators.

We note that the dimensions of these algebras (meaning the dimensions of the corresponding vector spaces) are $n(n+1)/2$ for the symmetric case and $n(n-1)/2$ for the antisymmetric case.

Biographical Sketches

Galois, Èvariste (1811–32) was born in Bourg-la-Reine, France. His mathematical genius was not recognized while he was alive. At the age of seventeen he submitted a paper for presentation to the Académie, which was lost by Cauchy; another paper submitted later was rejected by Poisson as "incomprehensible". He was killed in a duel. On the night before the duel he spent hours jotting down notes for posterity concerning his discoveries. He used group theory to show when an algebraic equation could be solved by radicals.

Abel, Niels Henrik (1802–29) was born in Finnøy, Norway. He lived in poverty and died of tuberculosis. Two days after his death a letter was written informing him that he was to be appointed professor of mathematics at the University of Berlin. He too had an important paper misplaced by Cauchy. In 1824 he published a proof that there is no algebraic formula for the roots of a general algebraic equation of fifth degree. He developed independently of Jacobi the theory of elliptic functions emphasizing their analogy with the familiar trigonometric functions. In his studies of the theory of equations he used commutative groups.

Jacobi, Carl Gustav Jacob (1804–51) was born in Potsdam, Germany. He published in 1829 *Fundamenta nova theoriae functionum ellipticarum*, the first definitive book on elliptic functions. He developed an approach to mechanics that we refer to today as the Hamilton–Jacobi formalism. The Jacobi identity was shown by him in application to the Poisson bracket, a structure isomorphic to the commutator.

2
Lie groups and Lie algebras

Lastly in our list of definitions we would like to get to the words Lie groups and Lie algebras. Suppose that we have a continuous group, where elements of the group can be parameterized by, say, d parameters, i.e. d real numbers that vary continuously: we have a **d-dimensional manifold**. Thus, at every point in parameter space we can erect a Cartesian coordinate system with d orthogonal axes. This means that statements can be made about the topology of the parameter space— which is what is meant by the topology of the group itself.

Let us consider here briefly the example of $SU(2)$: the group whose elements are unitary unimodular 2×2 matrices with complex entries and for which the composition law is ordinary matrix multiplication. A word about the name: $U(n)$ denotes the group whose elements are $n \times n$ unitary matrices, $SU(n)$—S for special—stands for the subgroup of $U(n)$ consisting of unimodular matrices. A matrix Z is **unitary** if its hermitian adjoint (i.e. complex conjugate transpose) is equal to its inverse:

$$Z^\dagger = Z^{-1} \qquad (2.1)$$

and it is **unimodular** if its determinant is equal to $+1$. We should verify that such matrices form a group. The only property that is not completely obvious is closure. If $Z, Y \in SU(2)$ then

$$(ZY)^\dagger = Y^\dagger Z^\dagger = Y^{-1} Z^{-1} = (ZY)^{-1} \quad \text{and} \quad \det(ZY) = (\det Z)(\det Y) = 1, \qquad (2.2)$$

so closure is obeyed.

An element of $SU(2)$ can be written as

$$Z = \begin{pmatrix} a & b \\ c & d \end{pmatrix}. \qquad (2.3)$$

Above is any 2×2 matrix with complex entries—it is obviously an 8-parameter quantity: the real and imaginary parts of a, b, c and d. If we write the unitarity statement, (2.1), as

$$1 = ZZ^\dagger \tag{2.4}$$

we have quite explicitly (the * denotes complex conjugate)

$$\begin{pmatrix} 1 & 0 \\ 0 & 1 \end{pmatrix} = \begin{pmatrix} a & b \\ c & d \end{pmatrix} \begin{pmatrix} a^* & c^* \\ b^* & d^* \end{pmatrix} = \begin{pmatrix} aa^* + bb^* & ac^* + bd^* \\ ca^* + db^* & cc^* + dd^* \end{pmatrix}, \tag{2.5}$$

that is the rows (and columns) are orthonormal complex 2-vectors:

$$|a|^2 + |b|^2 = 1,$$
$$|c|^2 + |d|^2 = 1,$$
$$ac^* + bd^* = 0, \tag{2.6}$$
$$ca^* + db^* = 0,$$

which provides *four* constraints on our eight parameters because the first two equations involve only real quantities, while the next two equations involve complex quantities, but are in fact each others complex conjugate. By taking the determinant of the unitarity statement $1 = ZZ^\dagger$ we get

$$|\det Z| = 1 \Rightarrow \det Z = \exp i\alpha, \tag{2.7}$$

i.e. det Z involves just one real number α, so the unimodularity constraint is only *one* more constraint. So we have in fact a 3-parameter group. The reader who recognizes $SU(2)$ as angular momentum is not surprised.

Another example familiar to physicists is the group of Lorentz transformations. This is a 6-parameter group—three Euler angles to specify a rotation in space, three components of velocity to specify a boost. One can show that the relevant matrix group is $O(3,1)$ whose elements are 4×4 orthogonal matrices, which are parameterized by $4\times 3/2 = 6$ parameters.

We define a **Lie group** to be a group, which has an infinite number of elements, where the elements are analytic functions of d parameters. We should like somehow to exhibit these parameters and this is where Lie algebras come in. Suppose that $g(x)$ is an element of the Lie group parameterized by the d parameters $x = (x_1, x_2, ..., x_d)$. If we agree that the identity element is parameterized by $x = 0$ then any element sufficiently near the identity can be expressed as

$$g(x) = \exp i x_a X_a, \tag{2.8}$$

where the x_a are our parameters and the X_a are called **infinitesimal generators**. From now on it is assumed that repeated indices are summed over the appropriate range (from 1 to d in the present case).

The essential properties of our group are in the X_a, not the x_a, which are mere parameters. The X_a are linearly independent operators of the same nature as the g (meaning that if the g are, say, $n \times n$ matrices, then so are the X_a). The $x_a X_a$, i.e. all the linear combinations of the X_a, form a d-dimensional vector space, with the X_a a basis in that space. This vector space becomes an algebra, the **Lie algebra**, when we define on it vector multiplication to be the commutator:

$$X_a \# X_b = X_a X_b - X_b X_a \equiv [X_a, X_b]. \tag{2.9}$$

It follows from these definitions that a Lie algebra is antisymmetric and has the derivative property.

Since the vector product defines a vector in our vector space that is spanned by the X_a, we must have

$$[X_a, X_b] = i f_{ab}{}^c X_c, \tag{2.10}$$

which is a relation of fundamental importance in the theory. The appearance of the imaginary unit i in (2.10) is correlated with the i in (2.8) and results in the generators being hermitian in unitary representations (which is viewed as an advantage by physicists):

$$g^\dagger = g^{-1} \Rightarrow X_a^\dagger = X_a. \tag{2.11}$$

It follows that the $f_{ab}{}^c$, called **structure constants**, are real numbers because the generators are hermitian. Further, it follows from (2.10) that

$$f_{ab}{}^c = -f_{ba}{}^c \tag{2.12}$$

and that when the generators are represented by matrices these matrices must be *traceless*. In view of the relation $\det(\exp M) = \exp(\mathrm{tr} M)$, where $\mathrm{tr} M$ stands for the trace of the matrix M, it follows that only unimodular group elements can be expressed in the form (2.8).

Since the Lie algebra has the derivative property we have

$$[[X_a, X_b], X_c] + [[X_b, X_c], X_a] + [[X_c, X_a], X_b] = 0 \tag{2.13}$$

and it follows from (2.10) and (2.13) that a bilinear expression in the structure constants vanishes, which can be stated as

$$[T_a, T_b] = i f_{ab}{}^c T_c, \tag{2.14}$$

where the T_a are matrices constructed out of the structure constants according to

$$(T_a)_{bc} = -i f_{ab}{}^c. \tag{2.15}$$

Thus, we discover a representation of the Lie algebra by $d \times d$ matrices formed from the structure constants—this is the so-called **adjoint** representation. Because it involves $d \times d$ matrices we say that the **dimension** of this representation is d. We shall see that representations of other dimensions are also possible. The number d, in its capacity as the number of parameters, therefore number of generators, is called the **order** (also the **dimension**) of the group.

As mentioned before, the generators provide a basis for the Lie algebra—we can change that basis and then we have different structure constants although we have, of course, the same algebra as before. Some of this arbitrariness can be removed by noting that we can form out of the structure constants the object g_{ab} by

$$g_{ab} = -f_{am}{}^n f_{bn}{}^m = \mathrm{tr}(T_a T_b), \tag{2.16}$$

which is a symmetric real matrix, known as **Cartan's metric tensor**. Such a matrix can always be diagonalized. So a basis can be found in which

$$\mathrm{tr}(T_a T_b) = k_a \delta_{ab}, \text{(no sum)}, \tag{2.17}$$

where the k_a, being eigenvalues of a real symmetric matrix, are real. Ideally, we would like to find a basis where $g_{ab} = c \delta_{ab}$ (where c is some convenient constant) but of course that is only possible if all the eigenvalues are non-zero and of the same sign.

It turns out that all the eigenvalues are non-zero if and only if the algebra is semisimple—this is known as Cartan's criterion for semisimplicity and before proving it we must first define "semisimple." A group is called **simple** if it contains no invariant subgroups besides itself, the identity and possible discrete subgroups; it is called **semisimple** if it contains no Abelian invariant subgroups besides the identity and possible discrete subgroups. A one-dimensional group is not semisimple since it contains an Abelian subgroup—namely itself. To avoid the embarrassment of a simple group not being semisimple we add to the definition of a simple group the requirement that it have more than one dimension.

To obtain the corresponding concepts for the algebra we observe that the Lie algebra Λ_H of a subgroup H of G is a **subalgebra** of the Lie algebra Λ_G of G. That means that if we denote the generators in Λ_G by X_a and those in Λ_H by X_α, the Latin indices ranging over some set I, the Greek indices over some set J, then a basis can be chosen such that J is a subset of I and the structure constants obey

$$f_{\alpha\beta}{}^c = 0 \quad \text{for} \quad c \notin J. \tag{2.18}$$

If H is an invariant subgroup then Λ_H is an **invariant** subalgebra, or an **ideal**, and the structure constants obey in addition

$$f_{ab}{}^c = 0 \quad \text{for} \quad c \notin J. \tag{2.19}$$

Further, Λ_H is called Abelian if H is Abelian. In that case the structure constants obviously all vanish:

$$f_{\alpha\beta}{}^\gamma = 0. \tag{2.20}$$

Lastly, the algebra Λ_G is called **simple** if it contains no proper ideals and is not one-dimensional, it is called **semisimple** if it contains no Abelian ideals except $\{0\}$.

Now, Cartan's criterion is: a Lie algebra is semisimple if, and only if,

$$\det g_{ab} \neq 0. \tag{2.21}$$

Let us show that the determinant vanishes if the Lie algebra *contains* an Abelian ideal. Using now Greek letters to denote the indices in the subset J referring to the Abelian ideal we have

$$g_{a\alpha} = -f_{am}{}^n f_{\alpha n}{}^m = -f_{a\mu}{}^n f_{\alpha n}{}^\mu \quad \text{because} \quad f_{\alpha n}{}^m = 0 \text{ for } m \notin J$$
$$= f_{\mu a}{}^n f_{\alpha n}{}^\mu = f_{\mu a}{}^\nu f_{\alpha \nu}{}^\mu = 0 \quad \text{because} \quad f_{\alpha \nu}{}^\mu = 0,$$

hence the column $b = \alpha$ of the matrix g_{ab} is 0, hence $\det g_{ab} = 0$.

One way to think about semisimplicity is this: one would like to think of the structure constants as codifying all the information about the algebra; clearly if they all vanish the structure constants have little to say about the set X_a—it follows that having an Abelian invariant subalgebra is a nuisance. Another way this nuisance manifests itself is in the fact that the Cartan metric has no inverse.

We note that the Cartan metric can be used to lower the superscript in the definition of the structure constants:

$$if_{abc} = if_{ab}{}^d g_{dc} = if_{ab}{}^d \ \mathrm{tr}(T_d T_c) = \mathrm{tr}\left([T_a, T_b] T_c\right), \tag{2.22}$$

which shows that f_{abc} is completely antisymmetric (since the trace of the product of two matrices is invariant under the exchange of the two matrices). Of course for semisimple groups subscripts can be raised using the inverse of the Cartan metric.

As mentioned before it would be nice if a basis could be found in which the Cartan metric is proportional to δ_{ab}, which is equivalent to the Cartan metric being positive definite. This turns out to be possible if the group is semisimple and compact. We shall say that a group is **compact** if the volume in parameter space is finite.

Consider again the example of the $SU(2)$ group. We can use (2.6) and unimodularity to solve for c and d:

$$c = -b^*, \quad d = a^*$$

and so we can express $Z \in SU(2)$ in terms of two complex parameter a and b as

$$Z = \begin{pmatrix} a & b \\ -b^* & a^* \end{pmatrix}, \tag{2.23}$$

where the two complex parameters are subject to the constraint

$$|a|^2 + |b|^2 = 1 \tag{2.24}$$

due to unimodularity. The explicit parameterization is completed by observing that (2.24) is satisfied by

$$a = e^{i\phi} \cos\vartheta, \quad b = e^{i\psi} \sin\vartheta, \tag{2.25}$$

$$0 \leq \vartheta \leq \tfrac{\pi}{2}, \quad 0 \leq \phi < 2\pi, \quad 0 \leq \psi < 2\pi,$$

which shows that the range of the parameters is bounded, hence volume in parameter space is finite.

Another way of thinking about $SU(2)$ being compact is to note that the parameterization above is really telling us that topologically $SU(2)$ is S^3—a unit 3-sphere. What we mean is this: we call the real points x_1 and x_2 for which

$$x_1^2+x_2^2\leq 1 \tag{2.26}$$

the unit disc $\mathbb{D}^2$—a two-dimensional manifold embedded in $\mathbb{R}^2$, the two-dimensional Euclidean space. The *boundary* of $\mathbb{D}^2$, defined by all points for which the equality holds in (2.26) is called a 1-sphere, $\mathbb{S}^1$, a one-dimensional manifold (the circumference of a circle):

$$\mathbb{S}^1=\partial\mathbb{D}^2. \tag{2.27}$$

Similarly

$$x_1^2+x_2^2+x_3^2=1 \tag{2.28}$$

defines $\mathbb{S}^2$, the 2-sphere, a two-dimensional manifold, the thing that in our ordinary three-dimensional world we call the surface of a ball $\mathbb{B}^3$—a three-dimensional manifold embedded in $\mathbb{R}^3$:

$$\mathbb{S}^2=\partial\mathbb{B}^3. \tag{2.30}$$

Next $\mathbb{S}^3$, the 3-sphere, is defined by

$$x_1^2+x_2^2+x_3^2+x_4^2=1, \tag{2.31}$$

a three-dimensional manifold, the boundary of the four-dimensional ball $\mathbb{B}^4$, embedded in $\mathbb{R}^4$.

Going back to $SU(2)$ we note that if we express the complex numbers a and b in terms of their real and imaginary parts as follows

$$a=x_1+i\,x_2 \quad \text{and} \quad b=x_3+i\,x_4 \tag{2.32}$$

then the constraint (2.24) becomes (2.31), i.e. topologically $SU(2)$ is $\mathbb{S}^3$.

It remains to remark that the volume V_N of an N-sphere $\mathbb{S}^N$ (i.e. the N-dimensional surface of a ball of unit radius in $N+1$ dimensions) is finite. It is given by the formula

$$V_N=2\pi^{\frac{N+1}{2}}/\Gamma\left(\frac{N+1}{2}\right). \tag{2.33}$$

Thus, we conclude that the $SU(2)$ group is compact. The corresponding Lie algebra is denoted by $su(2)$—we follow the convention of denoting the group by upper case and the algebra by lower case letters. The generators of $su(2)$ must be 2×2 hermitian traceless matrices. Clearly the following four matrices

$$\begin{pmatrix} 1 & 0 \\ 0 & 0 \end{pmatrix} \begin{pmatrix} 0 & 0 \\ 0 & 1 \end{pmatrix} \begin{pmatrix} 0 & 1 \\ 0 & 0 \end{pmatrix} \begin{pmatrix} 0 & 0 \\ 1 & 0 \end{pmatrix}$$

span the space of 2×2 matrices. An equally good set is obtained by replacing the first two by half their sum and difference and the last two by half their sum and difference:

$$\frac{1}{2}\begin{pmatrix} 1 & 0 \\ 0 & 1 \end{pmatrix}, \quad \frac{1}{2}\begin{pmatrix} 1 & 0 \\ 0 & -1 \end{pmatrix}, \quad \frac{1}{2}\begin{pmatrix} 0 & 1 \\ 1 & 0 \end{pmatrix}, \quad \frac{1}{2}\begin{pmatrix} 0 & -1 \\ 1 & 0 \end{pmatrix}$$

and if we omit the first one (proportional to the unit matrix) the remaining three are a basis for 2×2 *traceless* matrices. Finally, if we multiply the last one by i we have a hermitian set and so we arrive at the following standard choice for the three linearly independent X_a:

$$X_1 = \frac{1}{2}\begin{pmatrix} 0 & 1 \\ 1 & 0 \end{pmatrix}, \quad X_2 = \frac{1}{2}\begin{pmatrix} 0 & -i \\ i & 0 \end{pmatrix}, \quad X_3 = \frac{1}{2}\begin{pmatrix} 1 & 0 \\ 0 & -1 \end{pmatrix} \qquad (2.34)$$

and therefore

$$[X_i, X_j] = i\varepsilon_{ijk} X_k, \qquad (2.35)$$

where the subscripts run over 1, 2 and 3 and where ε_{ijk} is antisymmetric under the exchange of any pair of subscripts and $\varepsilon_{123} = +1$. So for this algebra and this choice of basis we obtain for the Cartan metric

$$g_{ij} = -\varepsilon_{ikm}\varepsilon_{jmk} = 2\delta_{ij}, \qquad (2.36)$$

i.e. for $su(2)$ the Cartan metric is positive definite.

But now to see what **non-compact** means we take another example: $SL(2,\mathbb{R})$—the special linear group in two dimensions over the reals. This is the group whose elements are unimodular 2×2 matrices with real entries. With multiplication defined as ordinary matrix multiplication this is obviously a group. Again obviously an arbitrary real 2×2 matrix is specified by four parameters (its four matrix elements) and the unimodularity constraint reduces this to three parameters—what are they?

Let us write the 2×2 matrix A with real entries in the peculiar way

$$A = \begin{pmatrix} \alpha_1 + \beta_2 & -\alpha_2 + \beta_1 \\ \alpha_2 + \beta_1 & \alpha_1 - \beta_2 \end{pmatrix}, \qquad (2.37)$$

which can always be done since we have four real parameters α_1, α_2, β_1, β_2 to match the four matrix elements of A. Now $A \in SL(2,\mathbb{R})$ if

$$1 = \det A = \alpha_1^2 - \beta_2^2 + \alpha_2^2 - \beta_1^2 = |\alpha|^2 - |\beta|^2, \qquad (2.38)$$

where α and β are the complex numbers

$$\alpha = \alpha_1 + i\alpha_2, \quad \beta = \beta_1 + i\beta_2. \qquad (2.39)$$

Equation (2.38) is satisfied by the following parameterization

$$\alpha = e^{i\phi}\cosh\chi, \quad \beta = e^{i\psi}\sinh\chi, \qquad (2.40)$$

$$0 \leq \chi < \infty, \quad 0 \leq \phi < 2\pi, \quad 0 \leq \psi < 2\pi.$$

Clearly the range of the parameters is *unbounded*. Equivalently, we should recognize that the constraint (2.38) defines the three-dimensional manifold $\mathbb{H}^3$ whose volume is infinite. Instead of the *sphere* of the $SU(2)$ case we have a *hyperboloid*—the group $SL(2,\mathbb{R})$ is non-compact.

We now look at the $sl(2,\mathbb{R})$ algebra. In view of the relation between group elements and generators as given by (2.8) we need for the generators a set of three traceless 2×2 matrices that are *pure imaginary*—we can simply take the generators we found for $su(2)$ and multiply the two real ones by i. Thus, we have for the generators of $sl(2,\mathbb{R})$

$$Y_1 = iX_1 = \frac{1}{2}\begin{pmatrix} 0 & i \\ i & 0 \end{pmatrix}, \quad Y_2 = X_2 = \frac{1}{2}\begin{pmatrix} 0 & -i \\ i & 0 \end{pmatrix}, \quad Y_3 = iX_3 = \frac{1}{2}\begin{pmatrix} i & 0 \\ 0 & -i \end{pmatrix}$$

$$(2.41)$$

and therefore

$$[Y_1, Y_2] = iY_3, \quad [Y_2, Y_3] = iY_1, \quad [Y_3, Y_1] = -i Y_2, \qquad (2.42)$$

which looks just like the commutators in the $su(2)$ case except for the minus sign in the third commutator. If we now evaluate the metric tensor g_{ij} for this case we find

$$g_{ij} = (-1)^i 2\delta_{ij}, \qquad (2.43)$$

i.e. in contrast to $su(2)$ the metric is *indefinite*.

We note in passing that whereas the $su(2)$ generators were hermitian only one of the $sl(2,\mathbb{R})$ generators is hermitian and two are anti-hermitian. Thus, whereas the representation of $SU(2)$ by 2×2 matrices is unitary, the representation of $SL(2,\mathbb{R})$ by 2×2 matrices is not. These are manifestations of the theorem that states that unitary representations of compact groups are finite-dimensional, whereas unitary representations of non-compact groups are infinite-dimensional. This will be shown later using the concept of the quadratic Casimir operator and at that point we will complete the proof that g_{ab} is positive definite if the group is compact.

Biographical Sketches

Euler, Leonhard (1707–83) was born in Basel, Switzerland. At the age of seventeen he received a master degree in philosophy from the University of Basel. In 1727 he started working at the Imperial Russian Academy of Science in St. Petersburg, became Professor of Physics in 1731 and eventually head of the Mathematics department. He spent the years from 1741 to 1766 at the Berlin Academy, then returned to St. Petersburg where he died in 1783. He was most prolific, having published (some posthumously) 866 items in spite of becoming blind. His name is associated among other things with the problem of the seven bridges of Köningsberg, the relation $V-E+F=2$ connecting the number of vertices, edges and faces of certain polyhedra, the product formula for the Riemann zeta function, the base e of the natural logarithms, the relation $e^{i\varphi}=\cos\varphi+i\sin\varphi$ and, its special case, the identity $e^{i\pi}+1=0$.

Lie, Sophus (1842–99) was born in Nordfjordeide, Norway. He was visiting France with Felix Klein when the Franco–Prussian war broke out in 1871. Klein, being German, had to leave France in a hurry, while Lie decided to go hiking in the Alps. Because of his poor command of the French language and his Nordic appearance he was arrested as a German spy and spent about a month in prison before frantic efforts by Darboux got him released. While studying contact transformations arising from partial differential equations he developed an extensive theory of continuous families of transformations, now known as Lie groups.

Cartan, Èlie Joseph (1869–1951) was born in Dolomieu, France, the son of a blacksmith. He was professor in Paris from 1912 to 1940. He reformulated Lie's work and, followed by Weyl, created the theory of Lie groups. Among his discoveries are the theory of spinors and the exterior differential calculus. His name is attached to many concepts—to name just a few in group theory: Cartan metric tensor, Cartan subalgebra, Cartan matrix. His thesis is often quoted when referring to the classification of simple Lie groups.

3

Rotations: $SO(3)$ and $SU(2)$

Now this is supposed to be a familiar subject and we shall use it to illustrate some of our abstract concepts.

Two dimensions. With conventional Cartesian coordinates we have for the vector from the origin to the point P=(x,y) the relations

$$x = r\cos\alpha, \quad y = r\sin\alpha, \tag{3.1}$$

where r is the norm (or length) of our vector

$$r = (x^2 + y^2)^{1/2} \tag{3.2}$$

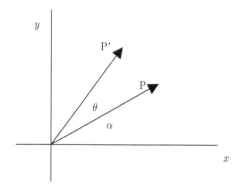

Fig. 3.1 *Rotation by an angle θ*

and α is the angle that the vector from the origin to P forms with the x-axis. As everybody knows if we rotate this vector by an angle θ counterclockwise the point P goes to P'=(x',y'), where

$$\begin{aligned} x' &= r\cos(\alpha+\theta) = r(\cos\alpha\cos\theta - \sin\alpha\sin\theta) = x\cos\theta - y\sin\theta \\ y' &= r\sin(\alpha+\theta) = r(\sin\alpha\cos\theta + \cos\alpha\sin\theta) = y\cos\theta + x\sin\theta. \end{aligned} \tag{3.3}$$

Clearly $x'^2+y'^2=r^2$, i.e. the rotation preserves the length of our vector. We may rewrite (3.3) as

$$\begin{pmatrix} x' \\ y' \end{pmatrix} = \begin{pmatrix} \cos\theta & -\sin\theta \\ \sin\theta & \cos\theta \end{pmatrix} \begin{pmatrix} x \\ y \end{pmatrix} \tag{3.4}$$

and view the 2×2 matrix

$$R(\theta) = \begin{pmatrix} \cos\theta & -\sin\theta \\ \sin\theta & \cos\theta \end{pmatrix} \tag{3.5}$$

as representing the rotation by θ counterclockwise.

Clearly we have a group because

$$R(\theta_1)R(\theta_2) = R(\theta_1+\theta_2) = R(\theta_2)R(\theta_1),$$
$$R(\theta)^{-1} = R(-\theta), \tag{3.6}$$
$$R(0) = 1_2$$

(where 1_2 denotes the 2×2 unit matrix). It is an Abelian one-parameter group, $0 \le \theta < \pi$, and it is compact. It is isomorphic to the additive group of real numbers mod 2π.

As we look at (3.5) we see that R is a 2×2 real matrix satisfying

$$R^{-1} = R^T \quad \text{and} \quad \det R = +1. \tag{3.7}$$

The first of these properties ensures that the length of our vector is left unchanged by the transformation and defines what is known as orthogonal matrices or transformations. We generalize to n dimensions by defining O to be an **orthogonal** matrix if it is an $n \times n$ real matrix satisfying

$$O^{-1} = O^T. \tag{3.8}$$

Writing this as $OO^T = 1$ and taking the determinant of that statement we have

$$1 = \det(OO^T) = (\det O)(\det O^T) = (\det O)^2$$

and we see that the determinant of an orthogonal matrix must be ± 1. When the determinant is $+1$ we have **rotations**, when it is -1 we have **reflections**. In our two-dimensional case an example of a transformation with determinant equal to -1 is

$$\begin{pmatrix} 1 & 0 \\ 0 & -1 \end{pmatrix} \qquad (3.9)$$

corresponding to reflecting one of the Cartesian axes. We note that the reflection of both Cartesian axes is described by

$$\begin{pmatrix} -1 & 0 \\ 0 & -1 \end{pmatrix}. \qquad (3.10)$$

This has determinant equal to $+1$ so is a rotation—comparing with (3.5) we see that it is a rotation by an angle $\theta=\pi$.

Clearly, orthogonal $n \times n$ matrices form a group called $O(n)$: the composition law is ordinary matrix multiplication, the identity is given by the unit matrix, the inverse is given by the transpose and closure holds because

$$(O_1 O_2)^{-1} = O_2^{-1} O_1^{-1} = O_2^T O_1^T = (O_1 O_2)^T. \qquad (3.11)$$

The unimodular orthogonal matrices, i.e. the rotations, form a subgroup called $SO(n)$. The reflections do *not* form a subgroup since they do not contain the identity nor obey closure.

Recalling the parameterization $g(x) = \exp i x_a X_a$, we see that $g^T = g^{-1}$ implies $X_a^T = -X_a$, i.e. the generators of the orthogonal group are given by *antisymmetric* matrices. Such matrices are of course traceless, as we already knew from (2.10), so the corresponding exponential is unimodular, that is we are dealing with $SO(n)$, the special orthogonal group. Here is an explicit example of the fact that only elements of the group connected with the identity element can be parameterized in our standard exponential fashion—a reflection cannot possibly be obtained by varying a parameter continuously from 0 (corresponding to the identity) to some finite value (corresponding to the reflection) because the value of the determinant cannot be *continuously* changed from $+1$ to -1 when those are the only values that the determinant can have.

We say that a space is **connected** if any two of its points can be joined by a curve lying in the space. It is clear from the preceding remarks that the orthogonal groups consist of two *disconnected* pieces corresponding to determinant $+1$ and -1. One of the two pieces is the one connected to the identity, the $SO(n)$ subgroup. The other piece is obtained by taking every element from the $SO(n)$ piece and multiplying it by, say, the diagonal $n \times n$ matrix

$$\mathrm{diag}(-1,1,1,...,1). \tag{3.12}$$

As we know the number of antisymmetric $n\times n$ matrices is $n(n-1)/2$, this is the number of generators and the number of parameters of $SO(n)$. Moreover, since the group elements are real the generators must be pure imaginary. Thus, in particular in two dimensions the $so(2)$ Lie algebra consists of just the one generator X of the form

$$X = \begin{pmatrix} 0 & i \\ -i & 0 \end{pmatrix}. \tag{3.13}$$

Any real multiple of X would still be an antisymmetric pure imaginary matrix but that the choice (3.13) is correct can be seen by the explicit calculation

$$\exp(i\theta X) = \sum_{n=0}^{\infty}(i\theta X)^n/n! = 1_2 \sum_{n=\text{even}}^{\infty} \theta^n/n! + iX \sum_{n=\text{odd}}^{\infty} \theta^n/n!$$
$$= 1_2\cos\theta + iX\sin\theta = R(\theta),$$

where we have used $(iX)^2 = 1_2$.

Three dimensions. Here things get to be a little more interesting. Clearly our two-dimensional results above can be viewed as rotating in some two-dimensional plane considered as a subspace of the familiar three-dimensional space that we live in. If we label the three orthogonal directions as 1, 2, 3 we have as the rotation by θ counterclockwise in the 1–2 plane

$$R_{12}(\theta) = \begin{pmatrix} \cos\theta & -\sin\theta & 0 \\ \sin\theta & \cos\theta & 0 \\ 0 & 0 & 1 \end{pmatrix} \tag{3.14}$$

—this is often referred to as a rotation by θ *about the axis 3*. In three dimensions the 1–2 plane can be associated with a single number—the missing number 3. In more than three dimensions this does not work and we must describe rotations as taking place in a plane rather than around an axis.

We saw that the group of rotations in two dimensions was Abelian. It follows that in three (and more) dimensions rotations in a given plane commute—but what about rotations in *different* planes? Of course everybody knows that the answer is no—rotations in different planes in three dimensions do not commute. In more than three dimensions it is possible for rotations in certain different planes to commute as we shall see in detail later.

Now, the group of transformations that leave the length of a vector in three dimensions unchanged is $O(3)$—the group whose elements are real 3×3 orthogonal matrices with multiplication defined as ordinary matrix multiplication. The subgroup of $O(3)$ consisting of unimodular matrices is $SO(3)$ and describes rotations, which can be parameterized in our standard manner with generators that must be 3×3 antisymmetric pure imaginary matrices. The number of antisymmetric 3×3 matrices is 3×2/2=3 and so a possible choice of our generators is

$$X_1 = i\begin{pmatrix} 0 & 0 & 0 \\ 0 & 0 & -1 \\ 0 & 1 & 0 \end{pmatrix}, \quad X_2 = i\begin{pmatrix} 0 & 0 & 1 \\ 0 & 0 & 0 \\ -1 & 0 & 0 \end{pmatrix}, \quad X_3 = i\begin{pmatrix} 0 & -1 & 0 \\ 1 & 0 & 0 \\ 0 & 0 & 0 \end{pmatrix} \quad (3.15)$$

so that

$$[X_i, X_j] = i\varepsilon_{ijk} X_k. \quad (3.16)$$

Since rotations in three dimensions are generated by **angular momentum** we shall refer to these X_i as the Cartesian components of the angular momentum operator in units such that the Plank constant h equals 2π.

Here, the ε_{ijk} is the same antisymmetric symbol that appeared in connection with the $su(2)$ Lie algebra and we have the remarkable result that the $su(2)$ and $so(3)$ Lie algebras are isomorphic. Comparing (3.15) with (3.13) we see that X_3 generates clockwise rotations in the 1–2 plane, X_1 in the 2–3 plane and X_2 in the 3–1 plane

We also see that the 3×3 matrices (3.15) can be written as

$$(X_i)_{jk} = -i\varepsilon_{ijk}, \quad (3.17)$$

in other words we have here the *adjoint* representation, which is three-dimensional.

The isomorphism of the $su(2)$ and $so(3)$ Lie algebras means that the $SU(2)$ and $SO(3)$ Lie groups are *locally* isomorphic, since the local properties of the group are completely described by the algebra. Not so for global properties. We have seen before that the topology of $SU(2)$ was that of $\mathbb{S}^3$, a 3-sphere. To obtain the topology of $SO(3)$ we remark that every rotation in three dimensions can be thought of as being about some axis by some angle no greater than π. So we can parameterize every rotation by a 3-vector whose direction is that of the axis of rotation and whose length is equal to the angle of rotation. Thus, if (ξ_1,ξ_2,ξ_3) specifies a point in parameter space of $SO(3)$ we have the requirement

$$\xi_1^2+\xi_2^2+\xi_3^2\leq\pi^2, \qquad (3.18)$$

which is $\mathbb{B}^3$, the three-dimensional ball. However, we now note that while every rotation by an angle smaller than π corresponds to a unique point *inside* this ball, rotations by π and $-\pi$ about any axis are the same rotation but correspond to diametrically opposite points on the *surface* of our ball. Thus, the parameter space of $SO(3)$ is $\mathbb{B}^3$ with opposite points on its surface identified.

Although the $SU(2)$ and $SO(3)$ groups are not isomorphic they are homomorphic, as we now explain. Let P=(y_1,y_2,y_3) and let P'=(y_1',y_2',y_3'), where P' is the point obtained from P by some rotation about the origin. We describe this rotation now in the following two ways:

i) Assemble the coordinates of P into a 3×1 (column) matrix Y, those of P' into Y' and write the relation between them as

$$Y'=RY, \qquad (3.19)$$

where the 3×3 matrix R describes the rotation. This is the standard approach and we know that $R\in SO(3)$.

ii) Assemble the coordinates of P into a 2×2 hermitian matrix H as follows

$$H=\begin{pmatrix} y_3 & y_1+iy_2 \\ y_1-iy_2 & -y_3 \end{pmatrix}. \qquad (3.20)$$

Similarly, assemble the coordinates of P' into H'. We have

$$\text{tr}H'=0=\text{tr}H,\ \det H'=-r'^2=-r^2=\det H, \qquad (3.21)$$

where the equality of the determinants is due to H' being obtained from H by a rotation. Therefore, we can relate H' to H by a similarity transformation

$$H'=QHQ^{-1} \qquad (3.22)$$

since such transformations preserve the trace and determinant. Since H and H' are complex 2×2 matrices so must be Q. Since H and H' are hermitian Q must be unitary (therefore Q^{-1} exists). But now we note that if Q satisfies (3.22) then so does cQ, c is an arbitrary number of unit magnitude. Without loss of generality we can restrict this arbitrariness by requiring det $Q=+1$. Thus, $Q \in SU(2)$ and $c^2=1$, i.e. $c=\pm 1$.

With these two ways of describing a rotation we arrive at the conclusion that R, an element of $SO(3)$, corresponds to the two elements Q and $-Q$ of $SU(2)$, i.e. the homomorphism of $SU(2)$ to $SO(3)$ is two-to-one.

Another way to describe the situation is to note that the two elements of $SU(2)$

$$\begin{pmatrix} 1 & 0 \\ 0 & 1 \end{pmatrix} \text{ and } \begin{pmatrix} -1 & 0 \\ 0 & -1 \end{pmatrix} \tag{3.23}$$

form a discrete invariant Abelian subgroup of $SU(2)$, isomorphic to $\mathbb{Z}_2$. Because these elements commute with every element of $SU(2)$ they form what is called the **center** of $SU(2)$. We may then form $SU(2)/\mathbb{Z}_2$, the so-called **factor group**, by identifying any two elements of $SU(2)$ whose ratio is an element from the center. This factor group *is* $SO(3)$ and so finally we arrive at the following isomorphism

$$SO(3) \cong SU(2)/\mathbb{Z}_2. \tag{3.24}$$

The parameter spaces of $SO(3)$ and $SU(2)$ also differ in their connectivity. As we said before a space is said to be connected if any two of its points can be joined by a curve lying in the space; we will say that it is **simply connected** if any two such curves connecting two points can be continuously deformed into one another. Equivalently, if the space is simply connected and has more than one dimension any loop can be shrunk to a point. All the $\mathbb{S}^n$, except for $\mathbb{S}^1$, are simply connected hence $SU(2)$ is simply connected since $\mathbb{S}^3$ is its group space. The group space of $SO(3)$—which is $\mathbb{S}^3/\mathbb{Z}_2$—is *not* simply connected. If we draw a curve in $\mathbb{S}^3$ from some point corresponding to the element u in $SU(2)$ to some point u' we cannot continuously deform it into a curve in $\mathbb{S}^3$ from u to $-u'$, even though these two curves join the same two points in $\mathbb{S}^3/\mathbb{Z}_2$. Alternately by thinking about the group space of $SO(3)$ as $\mathbb{B}^3$ with diametrically opposite points on the surface identified we can see that there are two types of loops: loops that can be shrunk to a point and loops that can be deformed to a diameter. We express that by saying that $SO(3)$ is *doubly connected*.

The picture below helps to explain why $\mathbb{S}^1$, in contrast to all the other $\mathbb{S}^n, n>1$, is not simply connected. Clearly the paths from a to b going clockwise or going counterclockwise cannot be deformed into each other if we are required to stay in $\mathbb{S}^1$, the circumference of the circle below. If, however, we are talking about, say, $\mathbb{S}^2$ with the picture below describing, say, the equator viewed from the north pole then the one path can easily be deformed into the other by sliding it over the northern hemisphere, all the deformations lying in $\mathbb{S}^2$.

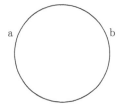

Fig. 3.2 *Connectivity*

There is an infinity of paths from a to b in $\mathbb{S}^1$ that cannot be deformed into each other, i.e. $\mathbb{S}^1$ is *infinitely connected*. We are referring here to paths from a to b winding around the center 0 times, 1 time, 2 times, etc. Recalling our parameterization of $SO(2)$ we see that the group space is $\mathbb{S}^1$, so $SO(2)$ is infinitely connected. As mentioned before, $SO(2)$ is isomorphic to the group of real numbers under addition mod 2π, i.e.

$$SO(2) \cong \mathbb{R}^1/\mathbb{Z}_\infty, \tag{3.25}$$

where we denote by $\mathbb{R}^1$ the additive group over the reals.

The isomorphism of the $so(3)$ and $su(2)$ Lie algebras raises the question: how many different groups can have the same Lie algebra? Let C denote a *simply connected* Lie group, let H denote its center, let $H_1, H_2,...$ denote subgroups of H. Then all the factor groups $G_k = C/H_k$, $k=1,2,...$ have the same Lie algebra as C but different connectivity properties. The group C is called the **universal covering group** of the groups G_k. As we have seen in the examples above the parameter space of C/Z_2 is doubly connected, that of C/Z_∞ is infinitely connected and in general the group H_k determines the connectivity properties of the parameter space of G_k.

4
Representations of $SU(2)$

In many ways the representation theory of $SU(2)$ typifies representation theory in general but of course is much simpler. In this chapter we will denote the three generators of $SU(2)$ by J_i, and take the commutation relations

$$[J_i, J_j] = i\varepsilon_{ijk} J_k, \tag{4.1}$$

(where i,j,k range over 1, 2, 3) as the definition of the $su(2)$ algebra. We say that we have a representation if we can represent the J_i by some matrices that satisfy (4.1). This representation will be a unitary representation of $SU(2)$ if the generators are hermitian:

$$J_i^\dagger = J_i. \tag{4.2}$$

We have already seen two representations: the representation in terms of 2×2 matrices, which is called the **defining** representation of $SU(2)$, and the representation in terms of 3×3 matrices, which we recall is the adjoint representation of $SU(2)$ [and the defining representation of $SO(3)$].

The above two representations could be glued together into a direct sum resulting in a 5×5 representation—such a representation is called **reducible**, in contrast to it our original 2×2 or 3×3 representations are called **irreducible** as they cannot be expressed as a direct sum of representations of lower dimensions. Clearly the task of finding all possible representations can be confined to finding all possible irreducible representations.

Let $|\gamma\rangle$ denote an eigenstate of J_3 to the eigenvalue γ:

$$J_3|\gamma\rangle = \gamma|\gamma\rangle, \quad \langle\gamma|\gamma\rangle \neq 0. \tag{4.3}$$

γ is real because J_3 is hermitian. We form out of J_1 and J_2 the two combinations

$$J_\pm = J_1 \pm i J_2 \tag{4.4}$$

and deduce from (4.1)
$$[J_3, J_\pm] = \pm J_\pm, \quad [J_+, J_-] = 2J_3. \tag{4.5}$$

Therefore,
$$J_3 J_\pm |\gamma\rangle = J_\pm |\gamma\rangle (\gamma \pm 1), \tag{4.6}$$

which means that either
$$J_\pm |\gamma\rangle = 0 \tag{4.7}$$

or else
$$J_\pm |\gamma\rangle \sim |\gamma \pm 1\rangle = \text{eigenstate of } J_3 \text{ to eigenvalue } \gamma \pm 1. \tag{4.8}$$

Thus, we find that J_+ can be viewed as a *raising* operator because when applied to an eigenstate of J_3 to some eigenvalue it either annihilates it or produces an eigenstate of J_3 with eigenvalue increased by 1. Similarly, J_- is a *lowering* operator.

Now, given the eigenvalue γ by repeatedly using these shift operators we can produce the string of eigenvalues
$$\ldots \gamma-3, \quad \gamma-2, \quad \gamma-1, \quad \gamma, \quad \gamma+1, \quad \gamma+2, \ldots . \tag{4.9}$$

We are interested, if possible, in finding *finite-dimensional* representations. That means that the above string must have a smallest and a largest term, i.e. there must exist *non-negative integers* a and b such that
$$J_+|\gamma+a\rangle = 0, \quad J_-|\gamma-b\rangle = 0, \tag{4.10}$$

i.e. $\gamma+a$ is the **highest** eigenvalue or **weight** of J_3 and $\gamma-b$ is the **lowest**.

So, let us start from the **highest weight state** $|j\rangle$ defined by
$$J_+|j\rangle = 0, \quad J_3|j\rangle = j|j\rangle, \quad \langle j|j\rangle = 1 \tag{4.11}$$

and ask for the effect of J_- on it. It follows from the discussion so far that J_- either annihilates the state or results in a state of weight $j-1$. In the latter case we can apply J_- again. Consider therefore applying J_- say, k times
$$(J_-)^k |j\rangle = N_k |j-k\rangle, \quad \langle j-k|j-k\rangle = 1, \tag{4.12}$$

where N_k is some number whose magnitude squared is the norm squared of the state $(J_-)^k|j\rangle$:

$$\begin{aligned}|N_k|^2 &= \langle j|(J_-)^{k\dagger}(J_-)^k|j\rangle = \langle j|(J_+)^k(J_-)^k|j\rangle \\ &= \langle j|(J_+)^{k-1}[J_+,(J_-)^k]|j\rangle \\ &= 2\langle j|(J_+)^{k-1}\{J_3(J_-)^{k-1}+J_-J_3(J_-)^{k-2}+...+(J_-)^{k-1}J_3\}|j\rangle \\ &= 2\langle j|(J_+)^{k-1}\{j-(k-1)+j-(k-2)+...+j\}(J_-)^{k-1}|j\rangle \\ &= k(2j-k+1)\langle j|(J_+)^{k-1}(J_-)^{k-1}|j\rangle \end{aligned} \quad (4.13)$$

Thus, we have the recursion relation

$$|N_k|^2 = k(2j-k+1)|N_{k-1}|^2 \quad (4.14)$$

with the solution

$$|N_k|^2 = \frac{k!(2j)!}{(2j-k)!}. \quad (4.15)$$

In order that the representation be finite-dimensional N_k must vanish for some k (which by construction is a positive integer) and we recognize the well-known angular momentum result that finite-dimensional irreducible representations can be labeled by j and exist for any j such that

$$2j = 0, 1, 2, ... \quad (4.16)$$

and we obtain a $(2j+1)$-dimensional representation.

We re-label $k = j - m$ and rewrite (4.12) as

$$|m\rangle = \frac{(J_-)^{j-m}}{N_{j-m}}|j\rangle$$

or

$$|m\rangle = \sqrt{\frac{(j+m)!}{(j-m)!(2j)!}}(J_-)^{j-m}|j\rangle, \quad -j \le m \le j, \quad (4.17)$$

where we have chosen the phase of N_{j-m} to be zero.

To find the explicit matrices that represent the generators we proceed as follows. Apply J_- to $|m\rangle$:

$$J_-|m\rangle = \sqrt{\frac{(j+m)!}{(j-m)!(2j)!}}(J_-)^{j-m+1}|j\rangle = \sqrt{(j+m)(j-m+1)}|m-1\rangle, \quad (4.18)$$

apply J_+ to $|m\rangle$:

$$J_+|m\rangle = [J_+, (J_-)^{j-m}]|j\rangle/N_{j-m} = (j-m)(j+m+1)(J_-)^{j-m-1}|j\rangle/N_{j-m}$$
$$= \sqrt{(j-m)(j+m+1)}|m+1\rangle, \quad (4.19)$$

that is our generators are represented by $(2j+1)\times(2j+1)$ matrices given by

$$(J_3)_{m'\,m} \equiv \langle m'|J_3|m\rangle = m\delta_{m',\,m} \quad (4.20)$$

$$(J_\pm)_{m'\,m} \equiv \langle m'|J_\pm|m\rangle = \sqrt{(j\mp m)(j\pm m+1)}\delta_{m',m\pm 1}. \quad (4.21)$$

In this derivation we have assumed that the representations should be finite-dimensional. In fact, this is true for unitary representations. To see this consider the following quadratic polynomial in the generators:

$$C_2 \equiv J_1^2 + J_2^2 + J_3^2$$
$$= J_3(J_3-1) + J_+J_- = J_3(J_3+1) + J_-J_+ \quad (4.22)$$

for which explicit calculation shows that

$$[C_2, J_i] = 0. \quad (4.23)$$

We note parenthetically that if we think of the J_i as components of a vector in three dimensions, as is implied by the commutation relations (4.1), then C_2 is the length squared of that vector, hence invariant under rotations, hence must commute with the generators of rotations.

It follows from (4.23) that the state $|\gamma\rangle$ defined by (4.3) as an eigenstate of J_3 can be taken to be simultaneously an eigenstate of C_2 to some eigenvalue c_2. Since moreover $[C_2, J_\pm] = 0$ it follows that all the other states in the string (4.9) are eigenstates of C_2 to the *same* eigenvalue c_2. Let β refer to some weight in the string (4.9). Then we have

$$\beta^2 = \langle \beta|J_3^2|\beta\rangle = \langle \beta|C_2 - J_1^2 - J_2^2|\beta\rangle$$
$$= c_2 - \langle \beta|J_1^2 + J_2^2|\beta\rangle \leq c_2, \quad (4.24)$$

where the last step follows because the J_i are hermitian in a unitary representation so the quantity being subtracted from c_2 is non-negative (and for the same reason c_2 is non-negative).

Thus, we find that the eigenvalue spectrum of J_3 is *bounded* and we have seen before that it is discrete. Consequently, unitary representations are finite-dimensional. We also note the relation between j and c_2 that follows from (4.22):

$$C_2|j\rangle = \{(J_3(J_3+1)+J_-J_+\}|j\rangle = j(j+1)|j\rangle. \tag{4.25}$$

In fact, since C_2 commutes with all the J_i we actually have $C_2|m\rangle = j(j+1)|m\rangle$. Thus, it is more appropriate to denote these states by $|j,m\rangle$ to emphasize the fact that they are simultaneous eigenstates of two operators.

The operator C_2 goes by the name of the quadratic Casimir operator and can be defined more generally for any Lie algebra. With the generators of the Lie algebra represented by matrices we can form polynomials out of the generators by multiplying the generators using ordinary matrix multiplication (in contrast to the antisymmetrized multiplication used in the definition of Lie algebras). Then a **Casimir operator** is a polynomial formed out of the generators, which commutes with all the generators. For a semisimple group the **quadratic** Casimir operator can be taken to be

$$C_2 = g^{ab} X_a X_b. \tag{4.26}$$

This is obviously a quadratic polynomial in the generators and it commutes with any generator because

$$[g^{ab} X_a X_b, X_c] = g^{ab}(if_{ac}{}^d X_d X_b + if_{bc}{}^d X_a X_d)$$
$$= ig^{ab} f_{ac}{}^d (X_d X_b + X_b X_d) = ig^{ab} g^{ed} f_{ace}\{X_d, X_b\} \tag{4.27}$$

and this vanishes because f_{ace}, which is antisymmetric in a and e, is being contracted with an expression that is symmetric in a and e.

Since in a unitary representation the generators X_a are hermitian the quadratic Casimir (4.26) will have positive eigenvalues in such a representation provided the Cartan metric is positive definite. Moreover, the eigenvalue of the quadratic Casimir provides a bound for the eigenvalues of any of the X_a that might be used to label the states in this representation. So we arrive at the conclusion that a positive definite Cartan metric implies that unitary representations are finite dimensional. Thus, group elements are represented by finite-dimensional unitary matrices, which means that the matrix elements are *bounded*, which means that the group is compact.

In contrast, suppose that for the Lie group G under consideration g^{ab} has some positive and some negative eigenvalues. Consider then another group G' obtained from G by multiplying all the generators associated with negative eigenvalues of g^{ab} by i (this is known as the Weyl unitary trick). For G' all the previous arguments apply and its unitary representations are finite-dimensional. These representations are also representations of the original group but *not unitary* since some generators are hermitian, some anti-hermitian; consequently the matrix elements are unbounded. We *can* form unitary representations, which then have bounded matrix elements, but these representations are infinite-dimensional. Thus, in this case the group is non-compact.

This completes the proof of the theorem hinted at in Chapter 2: a necessary and sufficient condition for the Cartan metric tensor to be positive definite is for the group to be semisimple and compact.

We might inquire whether we have found representations of $SU(2)$ or $SO(3)$. Since the derivation involved just the Lie algebra it follows that we have found representations of the covering group, i.e. of $SU(2)$, but some of these representations could fail to be representations of $SO(3)$. In particular, consider a rotation in the 1–2 plane by 2π:

$$\exp(i2\pi J_3)|j,m\rangle = \exp(i2\pi m)|j,m\rangle = (-1)^{2m}|j,m\rangle \qquad (4.28)$$

and it follows from the definition of m that $2m$ is even (odd) in a representation for which $2j$ is even (odd). Consequently, only the representations for which $2j$ is *even* correspond to representations of $SO(3)$ for which we demand that rotations by 2π should be the identity. (Note that for $2j$ odd we must rotate by 4π to reach the identity.)

We shall see later that the present result extends to rotations in n dimensions: the representations of $Spin(n)$, the covering group of $SO(n)$, come in two varieties labeled roughly speaking by integers or half-odd-integers (also called tensor representations and spinor representations) and only the integer-labeled (tensor) are representations of $SO(n)$.

4.1 Addition of angular momentum

Suppose that we have two independent angular momentum operators J_{ai} and J_{bi}, $i=1,2,3$, that act in two independent spaces. Each space is spanned by the states discussed above and we distinguish them by using the subscripts a and b, i.e. one space is spanned by the states $|j_a,m_a\rangle$, the other by the states $|j_b,m_b\rangle$. Consider next the space spanned by the product states $|j_a,m_a\rangle|j_b,m_b\rangle$. Operators in this space are of the form AB and their action is defined by

$$(AB)(|j_a,m_a\rangle|j_b,m_b\rangle)=(A|j_a,m_a\rangle)(B|j_b,m_b\rangle). \qquad (4.29)$$

Next, we define an operator J_i by

$$J_i = J_{ai}\mathbf{1} + \mathbf{1}J_{bi} \qquad (4.30)$$

(where we denote by $\mathbf{1}$ the unit operator in any space) and we have

$$\begin{aligned}
[J_i,J_j] &= (J_{ai}\mathbf{1}+\mathbf{1}J_{bi})(J_{aj}\mathbf{1}+\mathbf{1}J_{bj})-i\Leftrightarrow j\\
&=(J_{ai}J_{aj})\mathbf{1}+J_{ai}J_{bj}+J_{aj}J_{bi}+\mathbf{1}(J_{bi}J_{bj})-i\Leftrightarrow j \qquad (4.31)\\
&=i\,\varepsilon_{ijk}(J_{ak}\mathbf{1}+\mathbf{1}J_{bk})=i\,\varepsilon_{ijk}J_k,
\end{aligned}$$

that is J_i is an angular momentum, we think of it as the sum of the a- and b-type angular momentum. Since J_i is an angular momentum the space on which it acts is spanned again by the same kind of states, which we now denote by $|j,m\rangle$. These states and the product states span the same space and so must be expressible in terms of each other.

We start with the product state $|j_a,j_a\rangle|j_b,j_b\rangle$. We have

$$\begin{aligned}
J_3|j_a,j_a\rangle|j_b,j_b\rangle &= (j_a+j_b)|j_a,j_a\rangle|j_b,j_b\rangle, \qquad (4.32)\\
J_+|j_a,j_a\rangle|j_b,j_b\rangle &= 0,
\end{aligned}$$

therefore (up to an arbitrary phase)

$$|j_a,j_a\rangle|j_b,j_b\rangle = |j_a+j_b,j_a+j_b\rangle. \qquad (4.33)$$

Applying J_- to both sides of (4.33) gives

$$\begin{aligned}
\sqrt{j_a}|j_a,j_a-1\rangle|j_b,j_b\rangle &+ \sqrt{j_b}|j_a,j_a\rangle|j_b,j_b-1\rangle\\
&= \sqrt{j_a+j_b}|j_a+j_b,j_a+j_b-1\rangle. \qquad (4.34)
\end{aligned}$$

Since the two product states on the left of (4.34) are each eigenstates of J_3 to the eigenvalue j_a+j_b-1 if we form out of them a combination orthogonal to (4.34) we will get the state proportional to $|j_a+j_b-1,j_a+j_b-1\rangle$:

$$\sqrt{j_b}|j_a,j_a-1\rangle|j_b,j_b\rangle - \sqrt{j_a}|j_a,j_a\rangle|j_b,j_b-1\rangle$$
$$= -\sqrt{j_a+j_b}|j_a+j_b-1,j_a+j_b-1\rangle, \qquad (4.35)$$

where the factor $-\sqrt{j_a+j_b}$ on the right is found by evaluating the norm of the state on the left and again choosing a phase convention common in the literature.

There are three product states that are eigenstates of J_3 to the eigenvalue j_a+j_b-2, namely

$$|j_a,j_a-2\rangle|j_b,j_b\rangle, \quad |j_a,j_a-1\rangle|j_b,j_b-1\rangle, \quad |j_a,j_a\rangle|j_b,j_b-2\rangle \qquad (4.36)$$

and appropriate linear combinations will form the states $|j,j_a+j_b-2\rangle$ with $j=j_a+j_b$ or $j=j_a+j_b-1$ or $j=j_a+j_b-2$. We obtain $j=j_a+j_b$ by applying J_- to (4.34):

$$\sqrt{j_a(2j_a-1)}|j_a,j_a-2\rangle|j_b,j_b\rangle + 2\sqrt{j_a j_b}|j_a,j_a-1\rangle|j_b,j_b-1\rangle$$
$$+\sqrt{j_b(2j_b-1)}|j_a,j_a\rangle|j_b,j_b-2\rangle$$
$$= \sqrt{(j_a+j_b)(2j_a+2j_b-1)}|j_a+j_b,j_a+j_b-2\rangle, \qquad (4.37)$$

we obtain $j=j_a+j_b-1$ by applying J_- to (4.35):

$$\sqrt{j_b(2j_a-1)}|j_a,j_a-2\rangle|j_b,j_b\rangle + (j_b-j_a)|j_a,j_a-1\rangle|j_b,j_b-1\rangle$$
$$-\sqrt{j_a(2j_b-1)}|j_a,j_a\rangle|j_b,j_b-2\rangle$$
$$= -\sqrt{(j_a+j_b)(j_a+j_b-1)}|j_a+j_b-1,j_a+j_b-2\rangle, \qquad (4.38)$$

and finally we obtain $j=j_a+j_b-2$ by forming a properly normalized and phased linear combination orthogonal to (4.37) and (4.38):

$$\sqrt{j_b(2j_b-1)}|j_a,j_a-2\rangle|j_b,j_b\rangle - \sqrt{(2j_a-1)(2j_b-1)}|j_a,j_a-1\rangle|j_b,j_b-1\rangle$$
$$+\sqrt{j_a(2j_a-1)}|j_a,j_a\rangle|j_b,j_b-2\rangle$$
$$=\sqrt{(j_a+j_b-1)(2j_a+2j_b-1)}|j_a+j_b-2,j_a+j_b-2\rangle. \quad (4.39)$$

It is clear how to continue this procedure and obtain states $|j,m\rangle$ with lower and lower values of j and m. Eventually the process stops when we arrive at $j_{\min}$, the lowest possible value of j. To determine $j_{\min}$ we note that the number of different product states is $(2j_a+1)(2j_b+1)$ and the number of the $|j,m\rangle$ states must be the same as they span the same space. Thus, we must have

$$(2j_a+1)(2j_b+1) = \sum_{j=j_{\min}}^{j_a+j_b} (2j+1)$$
$$= (j_a+j_b-j_{\min}+1)(j_a+j_b+j_{\min}+1), \quad (4.40)$$

which yields

$$j_{\min} = |j_a-j_b|. \quad (4.41)$$

Because the product states are orthonormal the identity **1** can be expressed in terms of them as

$$\mathbf{1} = \sum_{m'_a,m'_b} |j_a,m'_a\rangle|j_b,m'_b\rangle\langle j_a,m'_a|\langle j_b,m'_b|, \quad (4.42)$$

where the sum over m'_a ranges from $-j_a$ to j_a in unit steps, and the sum over m'_b ranges from $-j_b$ to j_b in unit steps. Similarly, we have

$$\mathbf{1} = \sum_{j',m'} |j',m'\rangle\langle j',m'|, \quad (4.43)$$

where the sum over j' ranges from $|j_a-j_b|$ to j_a+j_b in unit steps, and the sum over m' ranges from $-j'$ to j' in unit steps. Therefore, we have

$$|j,m\rangle = \sum_{m'_a,m'_b} |j_a,m'_a\rangle|j_b,m'_b\rangle\langle j_a,m'_a|\langle j_b,m'_b|j,m\rangle$$
$$= \sum_{m'_a,m'_b} |j_a,m'_a\rangle|j_b,m'_b\rangle\langle j_a m'_a,j_b m'_b|jm\rangle, \quad (4.44)$$

where we have defined the so-called **Clebsch–Gordan coefficients**

$$\langle j_a m'_a, j_b m'_b | jm\rangle = \langle j_a, m'_a | \langle j_b, m'_b | j,m\rangle. \tag{4.45}$$

Similarly,

$$|j_a,m_a\rangle|j_b,m_b\rangle = \sum_{j',m'} |j',m'\rangle\langle j',m'|j_a,m_a\rangle|j_b,m_b\rangle$$

$$= \sum_{j',m'} |j',m'\rangle\langle j'm'|j_a m_a, j_b m_b\rangle, \tag{4.46}$$

where

$$\langle j'm'|j_a m_a, j_b m_b\rangle = \langle j_a m_a, j_b m_b|j'm'\rangle^*. \tag{4.47}$$

We note that our procedure given by (4.33)–(4.39) resulted in explicit values for several of these Clebsch–Gordan coefficients:

$$\langle j_a j_a, j_b j_b | j_a+j_b, j_a+j_b\rangle = 1,$$

$$\langle j_a j_a-1, j_b j_b | j_a+j_b, j_a+j_b-1\rangle = \sqrt{\frac{j_a}{j_a+j_b}},$$

$$\langle j_a j_a, j_b j_b-1 | j_a+j_b, j_a+j_b-1\rangle = \sqrt{\frac{j_b}{j_a+j_b}},$$

$$\langle j_a j_a-1, j_b j_b | j_a+j_b-1, j_a+j_b-1\rangle = -\sqrt{\frac{j_b}{j_a+j_b}}, \tag{4.48}$$

$$\langle j_a j_a, j_b j_b-1 | j_a+j_b-1, j_a+j_b-1\rangle = \sqrt{\frac{j_a}{j_a+j_b}},$$

$$\langle j_a j_a-2, j_b j_b | j_a+j_b, j_a+j_b-2\rangle = \sqrt{\frac{j_a(2j_a-1)}{(j_a+j_b)(2j_a+2j_b-1)}},$$

and so forth. We also note that our choice of phases in (4.33, 35, 39) is equivalent to a choice of phases for the Clebsch–Gordan coefficients and can be summarized in the statement: all Clebsch–Gordan coefficients are real and $\langle j_a j_a, j_b j-j_a | j,j\rangle$ is positive for all allowed j_a, j_b and j.

Equation (4.46) describes the so-called **Clebsch–Gordan series** of irreducible representations that result from the reduction of the product of two irreducible representations of $SU(2)$. We note that the Clebsch–Gordan coefficient $\langle j_a m'_a, j_b m'_b | jm\rangle$ vanishes except if

$$m'_a + m'_b = m \quad \text{and} \quad |j_a - j_b| \leq j \leq j_a + j_b \tag{4.49}$$

Table 4.1 $\langle j_a 1/2\, m_a\, m_b | j\, m\rangle$

j \ m_b	$1/2$	$-1/2$
$j_a+1/2$	$\sqrt{\frac{(j_a+m+\tfrac12)}{2j_a+1}}$	$\sqrt{\frac{(j_a-m+\tfrac12)}{2j_a+1}}$
$j_a-1/2$	$-\sqrt{\frac{(j_a-m+\tfrac12)}{2j_a+1}}$	$\sqrt{\frac{(j_a+m+\tfrac12)}{2j_a+1}}$

Table 4.2 $\langle j_a 1\, m_a\, m_b | j\, m\rangle$

j \ m_b	1	0	-1
j_a+1	$\sqrt{\frac{(j_a+m)(j_a+m+1)}{(2j_a+1)(2j_a+2)}}$	$\sqrt{\frac{(j_a-m+1)(j_a+m+1)}{(2j_a+1)(j_a+1)}}$	$\sqrt{\frac{(j_a-m)(j_a-m+1)}{(2j_a+1)(2j_a+2)}}$
j_a	$-\sqrt{\frac{(j_a+m)(j_a-m+1)}{2j_a(j_a+1)}}$	$\frac{m}{\sqrt{j_a(j_a+1)}}$	$\sqrt{\frac{(j_a-m)(j_a+m+1)}{2j_a(j_a+1)}}$
j_a-1	$\sqrt{\frac{(j_a-m)(j_a-m+1)}{2j_a(2j_a+1)}}$	$-\sqrt{\frac{(j_a-m)(j_a+m)}{j_a(2j_a+1)}}$	$\sqrt{\frac{(j_a+m+1)(j_a+m)}{2j_a(2j_a+1)}}$

and therefore the double sums in (4.44) and (4.46) actually collapse to a single sum.

The case when one of the angular momenta being added is 1/2 or 1 occurs frequently; the appropriate Clebsch–Gordan coefficients are given in the tables above.

4.2 The Wigner–Eckart theorem

It turns out that many operators of interest to Physicists behave under rotations in a manner similar to the states $|j,m\rangle$ discussed in this chapter. We recall that the $|j,m\rangle$ satisfy

$$J_3|j,m\rangle=m\,|j,m\rangle,$$
$$J_\pm|j,m\rangle=\sqrt{(j\mp m)(j\pm m+1)}|j,m\pm 1\rangle, \quad (4.50)$$
$$-j\le m\le j, \quad 2j=0,1,2,....$$

Now, an irreducible **spherical tensor** of rank s has $2s+1$ components denoted by O^s_λ, which satisfy the following commutation relations:

$$[J_3,O^s_\lambda]=\lambda O^s_\lambda,$$
$$[J_\pm,O^s_\lambda]=\sqrt{(s\mp\lambda)(s\pm\lambda+1)}O^s_{\lambda\pm 1}, \quad (4.51)$$
$$-s\le\lambda\le s, \quad s=0,1,2,....$$

It is these commutation relations that we have in mind when we say that a rank s irreducible tensor transforms under rotations like spin s. As an example, consider a vector consisting of the three Cartesian components V_i, $i=1,2,3$. We say that it is a vector because its commutation relations with the generators of rotations are

$$[J_i,V_j]=i\varepsilon_{ijk}V_k. \quad (4.52)$$

If we define the following linear combinations appropriate to the spherical basis

$$V_0=V_3, V_\pm=\mp(V_1\pm iV_2)/\sqrt{2} \quad (4.53)$$

then it follows from (4.52) that

$$[J_3,V_\lambda]=\lambda V_\lambda,$$
$$[J_\pm,V_\lambda]=\sqrt{(1\mp\lambda)(2\pm\lambda)}V_{\lambda\pm 1}, \quad (4.54)$$
$$-1\le\lambda\le 1,$$

so that a vector is in fact an irreducible spherical tensor of rank one, behaves like spin one. Examples would be the angular momentum itself, linear momentum, electric dipole moment, magnetic dipole moment, etc.

Consider next the state $O^s_s|j,j\rangle$. It satisfies

$$J_3 O^s_s|j,j\rangle=(j+s)O^s_s|j,j\rangle,$$
$$J_+ O^s_s|j,j\rangle=0, \quad (4.55)$$

and therefore must be proportional to $|j+s,j+s\rangle$:

$$O_s^s|j,j\rangle=|j+s,\ j+s\rangle o_{j+s,j}, \qquad (4.56)$$

where $o_{j+s,j}$ is some number that depends on j, s and on the nature of the operator O.

Acting on both sides of (4.56) with J_- we obtain

$$O_{s-1}^s|j,j\rangle\sqrt{s}+O_s^s|j,j-1\rangle\sqrt{j}=|j+s,j+s-1\rangle\sqrt{j+s}\,o_{j+s,j}. \qquad (4.57)$$

We note that both $O_{s-1}^s|j,j\rangle$ and $O_s^s|j,j-1\rangle$ are eigenstates of J_3 to the eigenvalue $j+s-1$ and (4.57) gives the linear combination proportional to $|j+s,j+s-1\rangle$, i.e. the combination that is an eigenstate of C_2 to the eigenvalue $(j+s)(j+s+1)$). There must exist another independent linear combination that is proportional to $|j+s-1,j+s-1\rangle$, i.e. an eigenstate of C_2 to the eigenvalue $(j+s-1)(j+s)$. Since

$$C_2 O_{s-1}^s|j,j\rangle$$
$$=O_s^s|j,j-1\rangle 2\sqrt{js}+O_{s-1}^s|j,j\rangle[(j+s-1)(j+s)+2s], \qquad (4.58)$$
$$C_2 O_s^s|j,j-1\rangle$$
$$=O_s^s|j,j-1\rangle[(j+s-1)(j+s)+2j]+O_{s-1}^s|j,j\rangle 2\sqrt{js} \qquad (4.59)$$

the desired combination is

$$O_{s-1}^s|j,j\rangle\sqrt{j}-O_s^s|j,j-1\rangle\sqrt{s}=-|j+s-1,j+s-1\rangle\sqrt{j+s}\,o_{j+s-1,j}, \qquad (4.60)$$

where $o_{j+s-1,j}$ is another number that depends on j, s and the nature of the operator O.

By now it should be clear how this procedure continues. But it should also be clear that this procedure is completely analogous to the procedure employed in the section on the addition of angular momentum. The point is that the state $O_\lambda^s|j,m\rangle$ is completely analogous to the product state $|s,\lambda\rangle|j,m\rangle$ with this one difference: whereas the product states are normalized, the norm squared

$$\langle j,m|O_\lambda^{s\dagger}O_\lambda^s|j,m\rangle \qquad (4.61)$$

is unknown—this is the explanation for the unknown quantities $o_{j+s,j}$, $o_{j+s-1,j}$, etc.

It then follows that we have the analog of (4.46)

$$O^s_\lambda |j,m\rangle = \sum_{j',m'} |j',m'\rangle \langle j'm'|s\lambda,jm\rangle o_{j',j} \qquad (4.62)$$

or equivalently

$$\langle j',m'|O^s_\lambda|j,m\rangle = \langle j'm'|s\lambda,jm\rangle o_{j',j}. \qquad (4.63)$$

Equation (4.63) is known as the Wigner–Eckart theorem. It states that the dependence of the matrix elements of our operator on the so-called magnetic quantum numbers m',m and λ is completely known as it is entirely contained in the Clebsch–Gordan coefficients. Thus, a rather large number of matrix elements is parameterized by the much smaller number of the different $o_{j',j}$, which are called **reduced matrix elements**.

All these ideas can be applied to groups other than $SU(2)$ or $SO(3)$. However, the Wigner–Eckart theorem is usually more powerful here because of a complication due to **multiplicity**. In general, in the reduction of the product of two irreducible representations a given irreducible representation may occur multiple times. If in the reduction of some particular two irreducible representations all representations occur either once or not at all, then this particular product is called **multiplicity-free**. If this is true for the products of *any* two irreducible representations than the group itself is called multiplicity-free. If the product is not multiplicity-free more reduced matrix elements appear, making the Wigner–Eckart theorem weaker. As we have seen, the $SU(2)$ and $SO(3)$ groups are multiplicity-free.

We remark parenthetically that we are talking here about *outer* multiplicity. Within a given irreducible representation a particular state can occur multiple times—this is known as *inner* multiplicity. The $SU(2)$ and $SO(3)$ groups are free of this multiplicity as well (in general these two multiplicities are related).

Biographical Sketches

Casimir, Hendrik Brugt Gerhard (1909–2000) was born in the Hague, Netherlands. He worked in numerous fields including the mathematical formalism of quantum mechanics. Beside the Casimir operator he has named after him the Casimir effect, which describes the difference in the zero-point energy of the electromagnetic field in the presence and absence of two condenser plates.

Weyl, Hermann (1885–1955) was born in Elmshorn, Germany. He studied under Hilbert in Göttingen, became professor at Zürich (1913) and Göttingen (1930). Refusing to stay in Nazi Germany he moved to Princeton in 1933. Inspired by a brief period with Einstein in Zürich he wrote on mathematical foundations of relativity and quantum mechanics. Subsequently, he wrote extensively on the representation theory of Lie groups. Modern gage theories can be traced back to Weyl's concept of measurement. The relativistic particle equation named after him describes a massless spin 1/2 particle. An obituary written by Freeman Dyson quotes Weyl as having said "My work always tried to unite the true with the beautiful; but when I had to choose one or the other, I usually chose the beautiful."

Clebsch, Rudolf Friedrich Alfred (1833–72) was born in Köningsberg, Germany (now Kaliningrad, Russia). He was Professor in Karlsruhe, Giessen and Göttingen. He did research on theory of invariants and application of elliptic theory and Abelian functions to geometry. He died in Göttingen.

Gordan, Paul (1837–1912) was born in Breslau, Germany (now Wroclaw, Poland). He was Professor in Giessen and Erlangen, and coauthored with Clebsch "Theory of Abelian Functions" in 1866. Gordan's only doctoral student, Emmy Noether, was one of the first women to receive a doctorate in Germany. He died in Erlangen.

Wigner, Eugene Paul (1902–95) was born in Budapest, Hungary. He spent four years at the famous Lutheran gymnasium of Budapest where he became friends with John von Neumann. In 1925 he received a degree in chemical engineering from the Technische Hohschule in Berlin. From 1930 to 1933 he spent part of the year in Berlin and part at Princeton University. His position in Berlin vanished when the Nazis came to power in 1933 and he joined the faculty of the new Institute for Advanced Study in Princeton. After three years there he went to the University of Wisconsin for two years and in the fall of 1938 went back to Princeton in an endowed professorship. At that time nuclear fission was discovered. With his friend Leo Szilard he persuaded Einstein to write a letter alerting Roosevelt to the dangers of the Nazis acquiring nuclear weapons. Among his many awards were: Medal for Merit, 1946; Fermi award, 1958; Max Planck Medal, 1961. In 1963 he shared the Nobel Prize for Physics with Maria Göppert-Mayer and J.H.D. Jensen for their contribution to the theory of nuclei and elementary particles, especially the discovery and application of principles of symmetry. One of his important contributions to group theory and physics was his construction of irreducible representations of the Poincaré group.

5

The $so(n)$ algebra and Clifford numbers

By definition, the elements of the $SO(n)$ group are real orthogonal unimodular $n \times n$ matrices with the composition law being ordinary matrix multiplication. Its generators are pure imaginary antisymmetric $n \times n$ matrices—so the *dimension* of $SO(n)$ is $n(n-1)/2$—and we can form a basis for the $so(n)$ algebra thus: name the $n(n-1)/2$ basis vectors $A_{ab} = -A_{ba}$, $a,b = 1,2,\ldots,n$. These basis vectors are $n \times n$ matrices with matrix elements given by

$$(A_{ab})_{st} = -\mathrm{i}(\delta_{as}\delta_{bt} - \delta_{at}\delta_{bs}) \equiv -\mathrm{i}\delta_{s[a}\delta_{b]t}, \tag{5.1}$$

i.e. matrices with $-\mathrm{i}$ at the intersection of the ath row and bth column, i at the transposed site and 0 everywhere else. Here, we introduce the convention that indices enclosed by square brackets are antisymmetrized.

It follows from (5.1) that

$$\begin{aligned}([A_{ij}, A_{mn}])_{st} &= (A_{ij})_{sp}(A_{mn})_{pt} - (ij \Leftrightarrow mn) \\ &= -\delta_{s[i}\delta_{j]p}\delta_{p[m}\delta_{n]t} - (ij \Leftrightarrow mn) \\ &= -\delta_{s[i}\delta_{j][m}\delta_{n]t} + \delta_{s[m}\delta_{n][i}\delta_{j]t} \\ &= -\mathrm{i}(A_{j[m}\delta_{n]i} - A_{i[m}\delta_{n]j})_{st}. \end{aligned} \tag{5.2}$$

We abstract from this result for the commutator in the defining representation and define the $so(n)$ algebra by

$$[A_{ij}, A_{mn}] = -\mathrm{i}(A_{j[m}\delta_{n]i} - A_{i[m}\delta_{n]j}), \tag{5.3}$$

where $A_{ij} = -A_{ji}$ and where in a unitary representation the generators are hermitian:

$$(A_{ij})^\dagger = A_{ij}. \tag{5.4}$$

If we rewrite (5.3) as

$$[A_{ij}, A_{mn}] = i f_{ij,mn}{}^{ks} A_{ks} \qquad (5.5)$$

then we see that the structure constants are given by

$$f_{ij,mn}{}^{ks} = \delta_{k[j}\delta_{i][m}\delta_{n]s}, \qquad (5.6)$$

hence we obtain for the Cartan metric tensor

$$g_{ij,ps} = -f_{ij,tu}{}^{km} f_{ps,km}{}^{tu}$$
$$= 2(n-2)\delta_{p[i}\delta_{j]s}\xrightarrow[i<j,p<s]{} 2(n-2)\delta_{ip}\delta_{js} \equiv 2(n-2)\delta_{ij,ps}, \qquad (5.7)$$

and conclude that $SO(n)$ is semisimple and compact for $n>2$.

A word about the constraint $i<j$, $p<s$. The pairs ij and ps in their capacity as labels of the rows and columns of the matrix $g_{ij,ps}$ should range over $n(n-1)/2$ values [corresponding to the number of linearly independent generators of $SO(n)$] and this can be achieved, without loss of generality, by requiring $i<j$, $p<s$. This constraint must be explicitly enforced since the pairs ij and ps, as they appear in the structure constants, range over $n(n-1)$ values, as only $i=j$, $p=s$ are excluded.

As mentioned before, the universal covering group of $SO(n)$, the so-called **Spin(n)** group, has two kinds of representations: spinor and tensor, of which only the latter are true representations of $SO(n)$ [one hears sometimes the spinor representations referred to as double-valued representations of $SO(n)$—that is an oxymoron]. The representation of the generators by $n \times n$ matrices as in (5.1) is, of course, the **defining** representation. This n-dimensional representation is also called the **vector** representation and is an example of a **tensor** representation, a vector being a tensor of rank one. By forming the direct product of two vector representations we obtain a rank two tensor representation, by forming a direct product of k vector representations we obtain a rank k tensor representation. This procedure results in *reducible* representations and we decompose them into irreducible components by symmetrizing or antisymmetrizing, or contracting with invariant tensors, as we shall see later. On the other hand, the *spinor* representations, discovered in 1913 by Cartan, cannot be obtained by this so-called tensoring of the vector representation.

We proceed to construct the spinor representation by way of Clifford numbers. Most physicists are familiar with Dirac's idea for extracting the square root of p^2, the square of the energy-momentum four-vector, by the following assumption:

$$p_1^2+p_2^2+p_3^2+p_4^2=(\gamma_1 p_1+\gamma_2 p_2+\gamma_3 p_3+\gamma_4 p_4)^2,$$

which works provided $\{\gamma_i,\gamma_j\}=2\delta_{ij}$, $i.j=1,2,3,4$.

Such entities were introduced in 1878 by Clifford. We define the **Clifford numbers** γ_i, $i=1, 2,..., n$, by the multiplication rule

$$\{\gamma_i, \gamma_j\}=2\delta_{ij}\mathbf{1}. \tag{5.8}$$

It follows from (5.8) that

$$[\gamma_i,\gamma_j\gamma_k]=\{\gamma_i,\gamma_j\}\gamma_k-\gamma_j\{\gamma_i,\gamma_k\}=2\delta_{i[j}\gamma_{k]}, \tag{5.9}$$

hence if we define

$$\Gamma_{ij}\equiv-\tfrac{i}{4}\gamma_{[i}\gamma_{j]} \tag{5.10}$$

then

$$\Gamma_{ij}=-\Gamma_{ji}, \tag{5.11}$$

$$[\Gamma_{ij},\Gamma_{mn}]=-i(\Gamma_{j[m}\delta_{n]i}-\Gamma_{i[m}\delta_{n]j}) \tag{5.12}$$

and comparison with (5.3) shows that the Γ_{ij} are a representation of the A_{ij}, the so-called **spinor** representation. This will be a unitary representation of $Spin(n)$ if the Clifford numbers are hermitian.

We note for future reference that with the definition (5.10) we can rewrite (5.9) as

$$[\Gamma_{ij}, \gamma_k]=-i\gamma_{[i}\delta_{j]k}, \tag{5.13}$$

which we will show means that under $so(n)$ rotations generated by the Γ_{ij} the Clifford numbers γ_k transform as the components of an n-vector. But first we continue with the discussion of the Clifford numbers.

Since

$$\gamma_i\gamma_j=\tfrac{1}{2}\{\gamma_i,\gamma_j\}+\tfrac{1}{2}[\gamma_i,\gamma_j]=\delta_{ij}\mathbf{1}+\tfrac{1}{2}\gamma_{[i}\gamma_{j]} \tag{5.14}$$

the space of the γs closes under multiplication if we include in addition to the $\mathbf{1}$ and the γs themselves all *completely antisymmetrized* products $\gamma_{[i}\gamma_{j]}$, $\gamma_{[i}\gamma_j\gamma_{k]}$, $\gamma_{[i}\gamma_j\gamma_k\gamma_{l]}$, etc. (or equivalently, products of distinct γs). Here, the notation means that subscripts enclosed by square brackets are to be completely antisymmetrized. So we find by requiring closure under multiplication that the following objects must be included:

Object	Number of different ones
1	$1 = \binom{n}{0}$
γ_i	$n = \binom{n}{1}$
$\gamma_{[i}\gamma_{j]}$	$n(n-1)/2 = \binom{n}{2}$
$\gamma_{[i}\gamma_j\gamma_{k]}$	$n(n-1)(n-2)/2\times 3 = \binom{n}{3}$
$\cdot$	$\cdot$
$\cdot$	$\cdot$
$\gamma_1\gamma_2\gamma_3\cdots\gamma_n$	$1 = \binom{n}{n}$,

(5.15)

where $\binom{n}{k}$ is the binomial coefficient; thus the total number of objects is

$$\sum_{k=0}^{n}\binom{n}{k} = 2^n. \tag{5.16}$$

For even n this is also the number of distinct $2^{n/2}\times 2^{n/2}$ matrices. Therefore, for $n=2m$ the Clifford numbers can be realized in terms of $2^m\times 2^m$ matrices, provided that we can display $2m$ such matrices obeying the defining relation (5.8). This we proceed to do by explicit construction.

We start by considering the simplest case, $m=1$, corresponding to 2×2 matrices. Setting

$$\rho = \begin{pmatrix} 0 & 1 \\ 1 & 0 \end{pmatrix}, \quad \sigma = \begin{pmatrix} 0 & -i \\ i & 0 \end{pmatrix} \tag{5.17}$$

it is trivial to verify that $\rho^2 = 1_2$, $\sigma^2 = 1_2$, $\rho\sigma = -\sigma\rho$, where 1_2 denotes the 2×2 unit matrix. Note that we are taking our matrices to be hermitian for future convenience.

For $m=2$ we are dealing with 4×4 matrices. We introduce the 2×2 matrix τ

$$\tau = \begin{pmatrix} 1 & 0 \\ 0 & -1 \end{pmatrix} \Rightarrow \tau = i\sigma\rho, \ \tau^2 = 1_2, \ \tau\rho = -\rho\tau, \ \tau\sigma = -\sigma\tau \tag{5.18}$$

and form the following 4×4 matrices:

$$\tau\otimes\rho, \quad \rho\otimes 1_2,$$
$$\tau\otimes\sigma, \quad \sigma\otimes 1_2, \tag{5.19}$$

where $\otimes$ means outer product so that the matrices (5.19) are 4×4. Quite explicitly we mean the following:

$$\tau\otimes\rho=\begin{pmatrix}\rho & 0_2 \\ 0_2 & -\rho\end{pmatrix}=\begin{pmatrix}0 & 1 & 0 & 0 \\ 1 & 0 & 0 & 0 \\ 0 & 0 & 0 & -1 \\ 0 & 0 & -1 & 0\end{pmatrix}, \quad \rho\otimes 1_2=\begin{pmatrix}0_2 & 1_2 \\ 1_2 & 0_2\end{pmatrix}=\begin{pmatrix}0 & 0 & 1 & 0 \\ 0 & 0 & 0 & 1 \\ 1 & 0 & 0 & 0 \\ 0 & 1 & 0 & 0\end{pmatrix}, \tag{5.20}$$

$$\tau\otimes\sigma=\begin{pmatrix}\sigma & 0_2 \\ 0_2 & -\sigma\end{pmatrix}=\begin{pmatrix}0 & -i & 0 & 0 \\ i & 0 & 0 & 0 \\ 0 & 0 & 0 & i \\ 0 & 0 & -i & 0\end{pmatrix}, \quad \sigma\otimes 1_2=\begin{pmatrix}0_2 & -i1_2 \\ i1_2 & 0_2\end{pmatrix}=\begin{pmatrix}0 & 0 & -i & 0 \\ 0 & 0 & 0 & -i \\ i & 0 & 0 & 0 \\ 0 & i & 0 & 0\end{pmatrix}. \tag{5.21}$$

Since matrix multiplication for two matrices written in this direct product form means

$$(A\otimes B)(C\otimes D)=(AC)\otimes(BD) \tag{5.22}$$

it is obvious that the four matrices (5.19) have unit square and mutually anticommute.

For $m=3$ we are dealing with 8×8 matrices and the following six are fine:

$$\tau\otimes\tau\otimes\rho, \quad \tau\otimes\rho\otimes 1_2, \quad \rho\otimes 1_2\otimes 1_2,$$
$$\tau\otimes\tau\otimes\sigma, \quad \tau\otimes\sigma\otimes 1_2, \quad \sigma\otimes 1_2\otimes 1_2. \tag{5.23}$$

The idea then, for any m, is as follows: for $s=1,2,...,m$ put

$$\rho_s=(\tau\otimes)^{s-1}\rho(\otimes 1_2)^{m-s},$$
$$\sigma_s=(\tau\otimes)^{s-1}\sigma(\otimes 1_2)^{m-s}. \tag{5.24}$$

It is then obvious that the $n=2m$ matrices (5.24) all mutually anticommute and have unit square. Setting then

$$\gamma_{2s-1}=\rho_s, \quad \gamma_{2s}=\sigma_s, \quad s=1,2,...,m \tag{5.25}$$

provides an explicit representation of the $n=2m$ Clifford numbers by $2^m \times 2^m$ matrices. This completes the proof of the isomorphism

$$C_{2m} \cong M(2^m), \tag{5.26}$$

where we denote by C_n the Clifford algebra with elements enumerated in (5.15) and by $M(k)$ the algebra of $k \times k$ matrices.

We now return to $so(n)$ and observe that we have a 2^m-dimensional representation in terms of the Γ_{ij}, the spinor representation, provided $n=2m$. This representation is unitary since in our explicit construction the γ_k were hermitian. However, as we now show, this representation is *reducible*.

As a preliminary we inquire into the commutation properties of the γ_k with a product of different γs. A product of an odd number of different γs commutes with every γ_k that is present in the product and anticommutes with those that are absent. A product of an even number of different γs anticommutes with every γ_k that is present in the product and commutes with those that are absent. In particular therefore, $\gamma_1 \gamma_2 \gamma_3 ... \gamma_{2m}$ anticommutes with every γ_k. Furthermore

$$\begin{aligned}(\gamma_1\gamma_2...\gamma_{2m})^2 &= \gamma_1\gamma_2...\gamma_{2m}\gamma_1\gamma_2...\gamma_{2m} \\ &= (-1)^{2m-1}\gamma_2...\gamma_{2m}\gamma_2...\gamma_{2m} \\ &= (-1)^{2m-1+2m-2}\gamma_3...\gamma_{2m}\gamma_3...\gamma_{2m} \\ &= \mathbf{1}(-1)^{\sum_{r=1}^{2m-1} r} = \mathbf{1}(-1)^{m(2m-1)} = \mathbf{1}(-1)^m. \end{aligned} \tag{5.27}$$

Therefore, with the definition

$$\gamma_{2m+1} \equiv (-i)^m \gamma_1 \gamma_2 ... \gamma_{2m} \tag{5.28}$$

we have

$$\{\gamma_i, \gamma_j\} = 2\delta_{ij}\mathbf{1}, \quad i, j = 1, 2, ..., 2m, 2m+1, \tag{5.29}$$

i.e. in the world of $2^m \times 2^m$ matrices there exist $2m+1$ entities that all anticommute and have unit square. We remark that when we use for the $2m$ γs the explicit representation given by (5.24) and (5.25) we have

$$\gamma_{2m+1} = \tau(\otimes \tau)^{m-1}. \tag{5.30}$$

We next observe that

$$[\Gamma_{ps}, \gamma_{2m+1}]=0, \tag{5.31}$$

where the indices p, s run over the $2m$ values corresponding to $so(2m)$. But if a matrix commutes with every generator in an irreducible representation then it must be proportional to the unit matrix—this is usually referred to as **Schur's lemma**. This means that for $so(2m)$ the 2^m-dimensional spinor representation is *reducible* and can be decomposed into so-called **semispinors**.

Matrices that satisfy the Clifford condition must have half of their eigenvalues equal to $+1$ and half to -1. (Since the squares of these matrices are the unit matrix, the squares of the eigenvalues must be 1; since any pair of these matrices anticommute they must be traceless, hence the number of $+1$ eigenvalues must equal the number of -1 eigenvalues.) So we can form the projection operators

$$P_\pm = \tfrac{1}{2}(\mathbf{1} \pm \gamma_{2m+1}), \tag{5.32}$$

which project onto the spaces where $\gamma_{2m+1}=\pm 1$; we can use these projection operators to reduce the 2^m-dimensional representation of $so(2m)$ into the two 2^{m-1}-dimensional irreducible semispinors. Moreover, the two semispinors are inequivalent since $\gamma_{2m+1}=+1$ in one representation, -1 in the other and there is no similarity transformation that can transform $+1$ into -1. This then completes the discussion of spinor representations of orthogonal algebras in *even* dimensions.

But of course we have also found a representation of the orthogonal algebra in *odd* dimensions—the Γ_{ij} provide a representation of $so(2m+1)$ if we let the indices range over the $2m+1$ values, i.e. include γ_{2m+1} in the list.

So we have a 2^m-dimensional spinor representation of $so(2m+1)$ and it is *irreducible* as (5.31) is no longer true when the indices p, s include the value $2m+1$.

To get a better understanding we look in detail at $so(2)$, $so(3)$ and $so(4)$. For $so(2)$ (i.e. $m=1$) the spinor representation is by 2×2 matrices and we have

$$\gamma_1 = \rho, \quad \gamma_2 = \sigma, \quad \Gamma_{12} = -\tfrac{i}{2}\gamma_1\gamma_2 = -\tfrac{i}{2}\rho\sigma = \tfrac{1}{2}\tau \tag{5.33}$$

and therefore the matrix representing counterclockwise rotation by θ is

$$\exp(-i\theta\Gamma_{12}) = \exp(-i\tfrac{\theta}{2}\tau) = 1_2\cos\tfrac{\theta}{2} - i\tau\sin\tfrac{\theta}{2} = \begin{pmatrix} \exp(-i\tfrac{\theta}{2}) & 0 \\ 0 & \exp(i\tfrac{\theta}{2}) \end{pmatrix}. \tag{5.34}$$

Such a diagonal matrix is typical of a reducible representation and we can reduce it to the two irreducible one-dimensional semispinors

$$\exp(-i\theta/2) \quad \text{and} \quad \exp(i\theta/2). \tag{5.35}$$

Since $\gamma_3 = -i\gamma_1\gamma_2 = \begin{pmatrix} 1 & 0 \\ 0 & -1 \end{pmatrix}$ we see that this reduction of the spinor is also reduction to the subspaces where $\gamma_3 = 1$ or $\gamma_3 = -1$. A rotation by 2π is represented by -1 for either semispinor and we see that we do not have representations of $SO(2)$, but rather representation of $Spin(2)$. We also note for future reference that the semispinors are complex and each others' complex conjugate.

This example is somewhat pathological since $SO(2)$ is one-dimensional and therefore Abelian and not semisimple.

Next we take up $so(3)$. The spinor representation is still 2×2 and the three γs are the same as given above. But now we have three generators as follows

$$\Gamma_{12} = -\tfrac{i}{2}\gamma_1\gamma_2 = -\tfrac{i}{2}\rho\sigma = -\tfrac{i}{2}\begin{pmatrix} 0 & 1 \\ 1 & 0 \end{pmatrix}\begin{pmatrix} 0 & -i \\ i & 0 \end{pmatrix} = \tfrac{1}{2}\begin{pmatrix} 1 & 0 \\ 0 & -1 \end{pmatrix},$$

$$\Gamma_{23} = -\tfrac{i}{2}\gamma_2\gamma_3 = -\tfrac{i}{2}\sigma\tau = -\tfrac{i}{2}\begin{pmatrix} 0 & -i \\ i & 0 \end{pmatrix}\begin{pmatrix} 1 & 0 \\ 0 & -1 \end{pmatrix} = \tfrac{1}{2}\begin{pmatrix} 0 & 1 \\ 1 & 0 \end{pmatrix},$$

$$\Gamma_{31} = -\tfrac{i}{2}\gamma_3\gamma_1 = -\tfrac{i}{2}\tau\rho = -\tfrac{i}{2}\begin{pmatrix} 1 & 0 \\ 0 & -1 \end{pmatrix}\begin{pmatrix} 0 & 1 \\ 1 & 0 \end{pmatrix} = \tfrac{1}{2}\begin{pmatrix} 0 & -i \\ i & 0 \end{pmatrix}, \tag{5.36}$$

which we recognize as the generators of $SU(2)$ as given in (2.34). Indeed $Spin(3)$ and $SU(2)$ are isomorphic.

Finally, consider $so(4)$. The four γs are given by (5.24) and (5.25) as

$$\gamma_1 = \rho \otimes 1_2, \quad \gamma_2 = \sigma \otimes 1_2, \quad \gamma_3 = \tau \otimes \rho, \quad \gamma_4 = \tau \otimes \sigma, \tag{5.37}$$

and therefore the six Γs are

$$\Gamma_{12}=-\tfrac{i}{2}\gamma_1\gamma_2=-\tfrac{i}{2}(\rho\otimes 1_2)(\sigma\otimes 1_2)=-\tfrac{i}{2}\rho\sigma\otimes 1_2=\tfrac{1}{2}\tau\otimes 1_2,$$

$$\Gamma_{13}=-\tfrac{i}{2}\gamma_1\gamma_3=-\tfrac{i}{2}(\rho\otimes 1_2)(\tau\otimes\rho)=-\tfrac{i}{2}\rho\tau\otimes\rho=-\tfrac{1}{2}\sigma\otimes\rho,$$

$$\Gamma_{14}=-\tfrac{i}{2}\gamma_1\gamma_4=-\tfrac{i}{2}(\rho\otimes 1_2)(\tau\otimes\sigma)=-\tfrac{i}{2}\rho\tau\otimes\sigma=-\tfrac{1}{2}\sigma\otimes\sigma,$$

$$\Gamma_{23}=-\tfrac{i}{2}\gamma_2\gamma_3=-\tfrac{i}{2}(\sigma\otimes 1_2)(\tau\otimes\rho)=-\tfrac{i}{2}\sigma\tau\otimes\rho=\tfrac{1}{2}\rho\otimes\rho, \quad (5.38)$$

$$\Gamma_{24}=-\tfrac{i}{2}\gamma_2\gamma_4=-\tfrac{i}{2}(\sigma\otimes 1_2)(\tau\otimes\sigma)=-\tfrac{i}{2}\sigma\tau\otimes\sigma=\tfrac{1}{2}\rho\otimes\sigma,$$

$$\Gamma_{34}=-\tfrac{i}{2}\gamma_3\gamma_4=-\tfrac{i}{2}(\tau\otimes\rho)(\tau\otimes\sigma)=-\tfrac{i}{2}1_2\otimes\rho\sigma=\tfrac{1}{2}1_2\otimes\tau.$$

We reduce this representation using the projection operators $P_\pm$, given by (5.32), to form

$$\Gamma^\pm_{ij}=P_\pm\Gamma_{ij}P_\pm=\tfrac{1}{2}(\Gamma_{ij}\pm\gamma_5\Gamma_{ij}). \qquad (5.39)$$

But now we find an unexpected bonus. In $so(4)$ we have $\gamma_5=-\gamma_1\gamma_2\gamma_3\gamma_4$ and therefore

$$\Gamma^\pm_{ij}=\tfrac{1}{2}(\Gamma_{ij}\pm\tfrac{1}{2}\varepsilon_{ijps}\Gamma_{ps}), \qquad (5.39)$$

which for convenience can be re-labeled as

$$\Gamma^\pm_\alpha=\tfrac{1}{2}\varepsilon_{\alpha\beta\gamma}\Gamma^\pm_{\beta\gamma}=\tfrac{1}{2}(\tfrac{1}{2}\varepsilon_{\alpha\beta\gamma}\Gamma_{\beta\gamma}\pm\Gamma_{\alpha 4}), \qquad (5.40)$$

where Latin indices range from 1 to 4, Greek indices from 1 to 3 and the εs are the appropriate antisymmetric symbols in 3 and 4 dimensions. What's remarkable is that these generators satisfy

$$[\Gamma^\pm_\alpha,\Gamma^\pm_\beta]=i\varepsilon_{\alpha\beta\gamma}\Gamma^\pm_\gamma, \quad [\Gamma^\pm_\alpha,\Gamma^\mp_\beta]=0, \qquad (5.41)$$

that is the "+" set and the "−" set not only commute with each other but also each set generates the algebra of $so(3)$—we have discovered the following algebra isomorphism

$$so(4)\cong so(3)\oplus so(3). \qquad (5.42)$$

Quite explicitly

$$\Gamma_1^+ = \tfrac{1}{2}(\Gamma_{23}+\Gamma_{14}) = \tfrac{1}{2}\begin{pmatrix} 0 & 0 & 0 & 1 \\ 0 & 0 & 0 & 0 \\ 0 & 0 & 0 & 0 \\ 1 & 0 & 0 & 0 \end{pmatrix},$$

$$\Gamma_2^+ = \tfrac{1}{2}(\Gamma_{31}+\Gamma_{24}) = \tfrac{1}{2}\begin{pmatrix} 0 & 0 & 0 & -i \\ 0 & 0 & 0 & 0 \\ 0 & 0 & 0 & 0 \\ i & 0 & 0 & 0 \end{pmatrix}, \qquad (5.43)$$

$$\Gamma_3^+ = \tfrac{1}{2}(\Gamma_{12}+\Gamma_{34}) = \tfrac{1}{2}\begin{pmatrix} 1 & 0 & 0 & 0 \\ 0 & 0 & 0 & 0 \\ 0 & 0 & 0 & 0 \\ 0 & 0 & 0 & -1 \end{pmatrix}.$$

If we ignore the second and third rows and columns (consisting of zeros only) we find precisely the 2×2 matrices (5.36) of the spinor representation of $so(3)$. For the "−" set the same procedure gives the same results except that now the first and fourth rows and columns must be ignored. This appearance of rows and columns of zeros is of course correlated with the following explicit form of γ_5:

$$\begin{pmatrix} 1 & 0 & 0 & 0 \\ 0 & -1 & 0 & 0 \\ 0 & 0 & -1 & 0 \\ 0 & 0 & 0 & 1 \end{pmatrix}. \qquad (5.44)$$

We rephrase these results as follows: in the subspace where $\gamma_5=1$ the six generators can be represented by the following 2×2 matrices:

$$\Gamma_1^+ = \tfrac{1}{2}\rho, \quad \Gamma_2^+ = \tfrac{1}{2}\sigma, \quad \Gamma_3^+ = \tfrac{1}{2}\tau, \quad \Gamma_\alpha^- = 0_2, \qquad (5.45)$$

while in the subspace where $\gamma_5=-1$ we have instead

$$\Gamma_\alpha^+ = 0_2, \quad \Gamma_1^- = \tfrac{1}{2}\rho, \quad \Gamma_2^- = \tfrac{1}{2}\sigma, \quad \Gamma_3^- = \tfrac{1}{2}\tau, \qquad (5.46)$$

these then are the two inequivalent two-dimensional semispinor representations of $so(4)$.

Biographical Sketches

Clifford, William Kingdon (1845–1879) was born in Exeter, England. He did research on non-Euclidean geometry, elliptic functions, and biquaternions. He was a first-class gymnast whose repertory supposedly included hanging by his toes from the crossbar of a weathercock on a church tower. He died of pulmonary tuberculosis in Madeira, Portugal.

Schur, Issai (1875–1941) was born in Mogilev, Russia (now Belarus). He became professor at Berlin in 1919, was forced to retire by the Nazis in 1935, was able to immigrate to Palestine in 1939 and died in Tel-Aviv of a heart ailment. He was a member of the Prussian Academy of Sciences before the Nazi purges. He did research on representation theory of groups, which was founded just before 1900 by his teacher Frobenius. Certain functions appearing in his work were named "S-functions" in his honor by English mathematicians.

6
Reality properties of spinors

Next we inquire into reality properties of the spinor representations of so(n).

For any algebra we say that we have a representation in terms of some matrices M_a provided the commutation relations

$$[M_a, M_b] = i f_{ab}{}^c M_c \tag{6.1}$$

are satisfied, where a, b, c range over the dimension of the algebra and $f_{ab}{}^c$ are its structure constants. If we take the transpose of (6.1) we see that the matrices $\overline{M}_a$, where

$$\overline{M}_a = -M_a^{\mathrm{T}}, \tag{6.2}$$

also satisfy (6.1), i.e. provide a representation. The two representations M_a and $\overline{M}_a$ are called each others' **conjugate** (also **contragradient**). A representation is called **self-conjugate** if it is similar to its conjugate, i.e.

$$-M_a^{\mathrm{T}} = C M_a C^{-1} \tag{6.3}$$

for some matrix C and for all M_a; otherwise the representation is called **complex**.

Let us demonstrate that (6.3) implies that C is either symmetric or antisymmetric. Taking the transpose of (6.3) gives

$$-M_a = (C^{-1})^{\mathrm{T}} M_a^{\mathrm{T}} C^{\mathrm{T}} \tag{6.4}$$

and re-expressing M_a^{T} by (6.3) again gives

$$M_a = (C^{-1})^{\mathrm{T}} C M_a C^{-1} C^{\mathrm{T}} \tag{6.5}$$

i.e.

$$M_a C^{-1} C^{\mathrm{T}} = C^{-1} C^{\mathrm{T}} M_a, \tag{6.6}$$

which shows that the matrix $C^{-1}C^{\mathrm{T}}$ commutes with all M_a and therefore, by Schur's lemma, must be (in an irreducible representation) proportional to the identity:

$$C^{-1}C^{\mathrm{T}}=\lambda \mathbf{1}. \tag{6.7}$$

Multiplying (6.7) from the left by C gives

$$C^{\mathrm{T}}=\lambda C, \tag{6.8}$$

hence

$$C=(\lambda C)^{\mathrm{T}}=\lambda C^{\mathrm{T}}=\lambda\lambda C \tag{6.9}$$

and therefore $\lambda=\pm 1$, i.e.

$$C^{\mathrm{T}}=\pm C. \tag{6.10}$$

So there are two kinds of self-conjugate representations—those for which C is *symmetric* are called **orthogonal** (also **real**), and those for which C is *antisymmetric* are called **symplectic** (also **pseudo-real**, also **quaternionic**).

The distinction between the two types has the following significance. Suppose that a similarity transformation can be found resulting in representation matrices that are antisymmetric. In other words a matrix U can be found such that

$$N_a^{\mathrm{T}}=-N_a, \tag{6.11}$$

where

$$N_a=UM_aU^{-1}. \tag{6.12}$$

Then

$$M_a^{\mathrm{T}}=(U^{-1}N_aU)^{\mathrm{T}}=-U^{\mathrm{T}}N_aU^{-1\mathrm{T}}=-U^{\mathrm{T}}(UM_aU^{-1})U^{-1\mathrm{T}}. \tag{6.13}$$

Comparing (6.13) and (6.3) we see that

$$C=U^{\mathrm{T}}U \tag{6.14}$$

and therefore C is *symmetric*.

Thus, representation matrices for a self-conjugate representation can be made antisymmetric, but only if C is symmetric; and self-conjugate representations exist that cannot be made antisymmetric, namely those for which C is antisymmetric, i.e. the symplectic ones.

For group elements that can be obtained from the algebra by exponentiation we have

$$\bar{g}=\exp(ix_a\overline{M}_a), \tag{6.15}$$

hence for self-conjugate representations with antisymmetric $\overline{M}_a$

$$\bar{g}^T=\bar{g}^{-1}, \tag{6.16}$$

which explains why such representations are called *orthogonal*. Moreover, if the representation is unitary then

$$\bar{g}^*=\bar{g}, \tag{6.17}$$

which explains the name *real*.

We are now ready to look at reality properties of spinors. We shall show that the spinor representations of $so(2m+1)$ are self-conjugate for all m, while the semispinor representations of $so(2m)$ are self-conjugate for m even, complex for m odd. In addition, we shall identify which of the self-conjugate representations are orthogonal and which are symplectic. We shall do this by explicitly displaying the matrix C in the representation in which the Clifford numbers are given by (5.24), (5.25) and (5.30):

$$\gamma_{2s-1}=(\tau\otimes)^{s-1}\rho(\otimes 1_2)^{m-s}, \quad \gamma_{2s}=(\tau\otimes)^{s-1}\sigma(\otimes 1_2)^{m-s}, \quad s=1,2,\ldots,m, \tag{6.18}$$

$$\gamma_{2m+1}=\tau(\otimes\tau)^{m-1}. \tag{6.19}$$

Let the matrix C be

$$\begin{aligned}
&m=1: & C&=\sigma & &\Rightarrow C^T=-C, \\
&m=2: & C&=\rho\otimes\sigma & &\Rightarrow C^T=-C, \\
&m=3: & C&=\sigma\otimes\rho\otimes\sigma & &\Rightarrow C^T=+C, \\
&m=4: & C&=\rho\otimes\sigma\otimes\rho\otimes\sigma & &\Rightarrow C^T=+C, \\
&\;\;\vdots \\
&m=2k: & C&=\rho\otimes(\sigma\otimes\rho\otimes)^{k-1}\sigma & &\Rightarrow C^T=(-1)^k C, \\
&m=2k+1: & C&=(\sigma\otimes\rho\otimes)^k\sigma & &\Rightarrow C^T=(-1)^{k+1}C.
\end{aligned} \tag{6.20}$$

The indicated behavior of C under transposition follows because ρ is symmetric and σ is antisymmetric.

By direct calculation we obtain

$$C\gamma_{2s-1}C^{-1}=(-1)^m\gamma_{2s-1}, \quad C\gamma_{2s}C^{-1}=(-1)^{m+1}\gamma_{2s}, \quad s=1,2,...,m, \tag{6.21}$$

$$C\gamma_{2m+1}C^{-1}=(-1)^m\gamma_{2m+1}, \tag{6.22}$$

and therefore, since the γs with odd subscripts are symmetric, with even subscripts are antisymmetric,

$$C\gamma_j C^{-1}=(-1)^m\gamma_j^{\mathrm{T}}, \quad j=1,2,...,2m,2m+1. \tag{6.23}$$

We therefore arrive at the conclusion that

$$C\Gamma_{ij}C^{-1}=-\tfrac{i}{4}C(\gamma_i\gamma_j-\gamma_j\gamma_i)C^{-1}=-\tfrac{i}{4}(\gamma_i^{\mathrm{T}}\gamma_j^{\mathrm{T}}-\gamma_j^{\mathrm{T}}\gamma_i^{\mathrm{T}})$$
$$=-\tfrac{i}{4}(\gamma_j\gamma_i-\gamma_i\gamma_j)^{\mathrm{T}}=-\Gamma_{ij}^{\mathrm{T}}, \quad i,j=1,2,...,2m,2m+1, \tag{6.24}$$

i.e. the spinor representations of $so(n)$ are self-conjugate. More precisely: the spinor representation of $so(2m+1)$ is self-conjugate and irreducible, the spinor representation of $so(2m)$ is self-conjugate and reducible. In view of (6.22) the projection operators $P_\pm=\tfrac{1}{2}(\mathbf{1}\pm\gamma_{2m+1})$ are invariant under the similarity transformation C for m even, and are transformed into each other for m odd. Hence, the semispinors of $so(2m)$ are self-conjugate for m even and complex and each others' conjugate for m odd. Taking into account the behavior of C under transposition we summarize the reality properties of spinors and semispinors in Table 6.1:

Table 6.1 Properties of spinors

algebra	reality properties	spinor dimension	semispinor dimension	center structure
$so(8k)$	orthogonal		2^{4k-1}	$\mathbb{Z}_2\times\mathbb{Z}_2$
$so(8k+1)$	orthogonal	2^{4k}		$\mathbb{Z}_2$
$so(8k+2)$	complex		2^{4k}	$\mathbb{Z}_4$
$so(8k+3)$	symplectic	2^{4k+1}		$\mathbb{Z}_2$
$so(8k+4)$	symplectic		2^{4k+1}	$\mathbb{Z}_2\times\mathbb{Z}_2$
$so(8k+5)$	symplectic	2^{4k+2}		$\mathbb{Z}_2$
$so(8k+6)$	complex		2^{4k+2}	$\mathbb{Z}_4$
$so(8k+7)$	orthogonal	2^{4k+3}		$\mathbb{Z}_2$

For future convenience we have included in the last column information about the center of the appropriate covering group. This will be discussed in Chapter 8.

We show in the next chapter that tensor representations can be obtained by tensoring spinor representations. It should be obvious that representations obtained from the reduction of the product of two symplectic or two orthogonal representations will be orthogonal, while representations obtained from the reduction of the product of a symplectic and orthogonal representation will be symplectic. Thus, we can conclude that *all* of the representations of $so(8k)$, $so(8k+1)$ and $so(8k+7)$ are orthogonal, while for $so(8k+3)$, $so(8k+4)$ and $so(8k+5)$ some representations are orthogonal and some are symplectic.

To conclude this chapter we give an example of a spinor that is symplectic—the 2-dimensional spinor of $so(3)$. The three matrices representing the generators are given in (5.36) as

$$\Gamma_{12}=\tfrac{1}{2}\begin{pmatrix}1 & 0\\ 0 & -1\end{pmatrix}, \quad \Gamma_{23}=\tfrac{1}{2}\begin{pmatrix}0 & 1\\ 1 & 0\end{pmatrix}, \quad \Gamma_{31}=\tfrac{1}{2}\begin{pmatrix}0 & -i\\ i & 0\end{pmatrix}.$$

The representation is self-conjugate because

$$C\Gamma_{ij}C^{-1}=-\Gamma_{ij}^T, \quad C=\sigma=\begin{pmatrix}0 & -i\\ i & 0\end{pmatrix} \tag{6.25}$$

however only one, namely Γ_{31}, is antisymmetric and a similarity transformation that would make all the Γ_{ij} antisymmetric does not exist. This is obvious since in the world of 2×2 matrices the number of independent antisymmetric matrices is $2(2-1)/2=1$. In agreement with the symplectic nature of this representation the matrix C is antisymmetric.

7
Clebsch–Gordan series for spinors

We have mentioned before that the commutation relations

$$[\Gamma_{ij}, \gamma_k]=-i\gamma_{[i}\delta_{j]k}, \quad i,j,k=1,2,\ldots,n, \tag{7.1}$$

mean that the Clifford numbers γ_k transform like the components of an n-vector under $so(n)$ rotations. Let us explain what we mean by this language.

It is easily seen that the defining commutation relations for $so(n)$ are satisfied by the following realization of the generators

$$A_{ab} \to L_{ab} = -ix_{[a}\partial_{b]}, \quad \partial_b \equiv \partial/\partial x_b. \tag{7.2}$$

We recognize L_{ab} as being proportional to the components of the angular momentum operator in n dimensions. Since x_a and ∂_b are components of n-vectors and

$$[L_{ab}, x_k] = -ix_{[a}\delta_{b]k}, \quad [L_{ab}, \partial_k] = -i\partial_{[a}\delta_{b]k} \tag{7.3}$$

the conclusion about γ_k follows.

We write out (7.1) in detail, including explicitly all matrix indices written as superscripts in Greek letters (as always, summation over repeated indices is understood):

$$\Gamma_{ij}^{\alpha\beta}\gamma_k^{\beta\rho} - \gamma_k^{\alpha\beta}\Gamma_{ij}^{\beta\rho} = -i\gamma_{[i}^{\alpha\rho}\delta_{j]k}, \tag{7.4}$$

or

$$\Gamma_{ij}^{\alpha\beta}\gamma_k^{\beta\rho} - (\Gamma_{ij}^T)^{\rho\beta}\gamma_k^{\alpha\beta} = -i\gamma_r^{\alpha\rho}\delta_{r[i}\delta_{j]k}, \tag{7.5}$$

where we made an obvious change in the second term and added an extra sum over r in the last term. The reason for doing that is that "$-i\delta_{r[i}\delta_{j]k}$" is precisely equal to $(A_{ij})_{rk}$, i.e. the generators of $so(n)$ in the defining or *vector* representation, see (5.1). To clarify matters we write

Clebsch–Gordan series for spinors

$$-i\delta_{r[i}\delta_{j]k} = (A^{\text{vector}}_{ij})^{rk}, \tag{7.6}$$

$$\Gamma^{\alpha\beta}_{ij} = (A^{\text{spinor}}_{ij})^{\alpha\beta} \tag{7.7}$$

so that (7.5) becomes

$$(A^{\text{spinor}}_{ij})^{\alpha\beta}\gamma^{\beta\rho}_{k} + \overline{(A^{\text{spinor}}_{ij})^{\rho\beta}}\gamma^{\alpha\beta}_{k} + (A^{\text{vector}}_{ij})^{kr}\gamma^{\alpha\rho}_{r} = 0, \tag{7.8}$$

where we use the bar to denote the conjugate (i.e. negative transpose) representation. Equation (7.8) justifies the following interpretation of the object $\gamma^{\alpha\beta}_{k}$, where the Latin index ranges over the *vector* range $1,2,\dots,2m+1$ and the Greek indices range over the *spinor* range $1,2,\dots,2^m$:

$\gamma^{\alpha\beta}_{k}$ transforms as a vector in the index k, as a spinor in the index α, and as a conjugate spinor in the index β.

Equivalently and more conveniently, by multiplying (7.1) by C^{-1} from the right and using (6.24) to get

$$\Gamma_{ij}C^{-1} = -C^{-1}\Gamma^{T}_{ij} \tag{7.9}$$

we obtain

$$(A^{\text{spinor}}_{ij})^{\alpha\beta}(\gamma_k C^{-1})^{\beta\rho} + (A^{\text{spinor}}_{ij})^{\rho\beta}(\gamma_k C^{-1})^{\alpha\beta} + (A^{\text{vector}}_{ij})^{kr}(\gamma_r C^{-1})^{\alpha\rho} = 0, \tag{7.10}$$

which we interpret as:

$(\gamma_k C^{-1})^{\alpha\beta}$ transforms as a vector in the index k, as a spinor in the index α, and as a spinor in the index β.

But we can do even better than that. It follows that the object $\gamma_k \gamma_s$ transforms as a vector in the index k and as a vector in the index s—this is what is meant by a tensor of rank two. In general, the product of, say, r γs transforms by definition as a tensor of rank r. Such a tensor can be reduced by imposing various symmetries under permutation of the γs and in particular the completely antisymmetric combinations $\gamma_{[k}\gamma_{s]}$, $\gamma_{[k}\gamma_s\gamma_{t]}$, etc. are (with one exception) *irreducible*. We can consider all these tensors at once by defining a symbol γ_A thus

$$A = 0: (\gamma_A)^{\alpha\beta} = (1)^{\alpha\beta},$$

$$A = 1: (\gamma_A)^{\alpha\beta} = (\gamma_k)^{\alpha\beta},$$

$$A = 2: (\gamma_A)^{\alpha\beta} = (\gamma_{[k}\gamma_{s]})^{\alpha\beta},$$

$$\vdots$$

$$A = m : (\gamma_A)^{\alpha\beta} = (\gamma_{\underbrace{[i}\gamma_j \ldots \gamma_{p]}})^{\alpha\beta}. \tag{7.11}$$
$$\phantom{A = m : (\gamma_A)^{\alpha\beta} = (\gamma_{[i}\gamma_j} }_{m}$$

It follows from our previous considerations that $(\gamma_A C^{-1})^{\alpha\beta}$ transforms as an antisymmetric tensor of rank A in the subscript and as a spinor in either of the superscripts. Hence, if we think of the $2^m \times 2^m$ matrix $(\ )^{\alpha\beta}$ as the direct product of two spinor representations of dimension 2^m the object $(\gamma_A C^{-1})^{\alpha\beta}$ can be thought of the *transformation matrix* between the space of antisymmetric tensors and the product space

$$\mathbf{2^m} \otimes \mathbf{2^m}, \tag{7.12}$$

where we denote the spinor representations in bold and by their dimension.

What spans this space? We are discussing $so(2m+1)$ and take $m=3$ as an example to work out the details. The Clifford numbers are $2m=6$ in number and the dimension of the spinor representation is $2^m=8$. To start out we make the table below. We use now for subscripts that range over just *six* values early letters of the Latin alphabet. The point is that the six γ_a do not a vector of $so(7)$ make. As we very well know we must include the γ_7 defined by the last line in the table below. Similarly, the 15 $\gamma_{[a}\gamma_{b]}$ together with the six $\gamma_7\gamma_f$ form the 21 components of the antisymmetric tensor of rank two in $so(7)$. Lastly, the 20 $\gamma_{[a}\gamma_b\gamma_{c]}$ together with the 15 $\gamma_7\gamma_{[e}\gamma_{f]}$ form the 35 components of the antisymmetric tensor of rank three in $so(7)$. So indeed the tensors γ_A, $A=0,1,2,3$ span the space having a total number of components equal to $1+7+21+35=64=8\cdot 8$.

Object	Number of distinct ones
1	1
γ_a, $a=1,2,\ldots,6$	6
$\gamma_{[a}\gamma_{b]}$	$6\times 5/2 = 15$
$\gamma_{[a}\gamma_b\gamma_{c]}$	$6\times 5\times 4/2\times 3 = 20$
$\gamma_{[a}\gamma_b\gamma_c\gamma_{d]} \sim \gamma_7\gamma_{[e}\gamma_{f]}$	15
$\gamma_{[a}\gamma_b\gamma_c\gamma_d\gamma_{e]} \sim \gamma_7\gamma_f$	6
$\gamma_1\gamma_2\gamma_3\gamma_4\gamma_5\gamma_6 = -i\gamma_7$	1

$$\tag{7.13}$$

Thus, we may summarize as follows the remarkable conclusion on the direct product of two spinor representations in $so(2m+1)$:

Clebsch–Gordan series for spinors

$$\mathbf{2}^m \otimes \mathbf{2}^m = \sum_{A=1}^{m} \oplus [\mathbf{A}], \qquad (7.14)$$

where we denote the spinor representations by their dimensions and the representation corresponding to the antisymmetric tensor of rank A by $[\mathbf{A}]$. Such a reduction of the direct product of two representations into the irreducible components goes under the name of the **Clebsch–Gordan series**.

We remark that since

$$\dim[\mathbf{A}] = \binom{2m+1}{A} \qquad (7.15)$$

we can verify as a final check that the dimensions on both sides of (7.14) are equal:

$$\sum_{A=0}^{m} \binom{2m+1}{A} = \tfrac{1}{2}\left(\sum_{A=1}^{m} + \sum_{A=m+1}^{2m+1}\right)\binom{2m+1}{A} = \tfrac{1}{2} 2^{2m+1} = 2^m \cdot 2^m. \qquad (7.16)$$

We recognize in the sequence

$$\mathbf{2}^m \otimes \mathbf{2}^m = [\mathbf{0}] \oplus [\mathbf{1}] \oplus [\mathbf{2}] \oplus \ldots \qquad (7.17)$$

the one-dimensional *trivial* representation, the $(2m+1)$-dimensional *vector* representation and the $((2m+1)m)$-dimensional *adjoint* representation. In particular, the occurrence of the vector representation means that all the representations that can be obtained by tensoring the vector representation can also be obtained by tensoring the spinor, i.e. *all* representations can be obtained by tensoring the spinor. We obtain so-called spin representations by tensoring an odd number of spinors, so-called tensor representations by tensoring an even number of spinors. Since the spinor of $Spin(2m+1)$ is self-conjugate, all the representations of $Spin(2m+1)$ are self-conjugate.

Next, we look into what happens for the *even* orthogonal algebras—here things turn out to be even more interesting. We can view $so(2m)$ as a subgroup of $so(2m+1)$, i.e. all the preceding considerations about the orthogonal group in *odd* dimensions apply except that we omit everywhere the value $2m+1$ from the range of the Latin indices. In particular this means that the γ_k, $k=1,2,\ldots,2m$ are to be viewed as the components of a $2m$-vector and the dimension of the antisymmetric tensor of rank A is given by

$$\dim[A] = \binom{2m}{A}. \tag{7.18}$$

Further, we recall that the $P_\pm = \frac{1}{2}(1 \pm \gamma_{2m+1})$ act as projection operators that can be used to reduce the 2^m-dimensional spinor of $so(2m)$ into the two 2^{m-1}-dimensional irreducible semispinors:

$$\mathbf{2^m} = (\mathbf{2^{m-1}})_+ \oplus (\mathbf{2^{m-1}})_-, \quad (\mathbf{2^{m-1}})_\pm = P_\pm(\mathbf{2^m}). \tag{7.19}$$

Since

$$\gamma_A C^{-1} = (P_+ + P_-)\gamma_A C^{-1}(P_+ + P_-) \tag{7.20}$$

we now have the *four* statements: $(P_u \gamma_A C^{-1} P_v)^{\alpha\beta}$ can be viewed as the transformation matrix between the space of antisymmetric tensors and the product space

$$(\mathbf{2^{m-1}})_u \otimes (\mathbf{2^{m-1}})_v, \quad u,v = \pm. \tag{7.21}$$

Recalling that γ_{2m+1} is symmetric, that it anticommutes with γ_k, $k=1,2,\ldots,2m$, and that $C\gamma_{2m+1}C^{-1} = (-1)^m \gamma_{2m+1}$, we see that

$$\text{for even } A: \quad P_\pm \gamma_A C^{-1} = \gamma_A P_\pm C^{-1} = \begin{cases} \gamma_A C^{-1} P_\pm & \text{for even } m \\ \gamma_A C^{-1} P_\mp & \text{for odd } m \end{cases} \tag{7.22}$$

$$\text{for odd } A: \quad P_\pm \gamma_A C^{-1} = \gamma_A P_\mp C^{-1} = \begin{cases} \gamma_A C^{-1} P_\mp & \text{for even } m \\ \gamma_A C^{-1} P_\pm & \text{for odd } m \end{cases}. \tag{7.23}$$

Lastly, since

$$P_\pm \gamma_{2m+1} = \gamma_{2m+1} P_\pm = \pm P_\pm \tag{7.24}$$

it follows that only the tensors for $A = 0,1,2,\ldots,m$ are independent. Moreover, the tensor of rank m is reducible, as we shall see in a moment.

Putting all these results together we arrive at the following Clebsch–Gordan series for the reduction of the product of two semispinors into antisymmetric tensors in $so(2m)$:

Clebsch–Gordan series for spinors

<u>m even</u>, say $m=2s$ so we are dealing with $so(4s)$:

$$(\mathbf{2^{m-1}})_\pm \otimes (\mathbf{2^{m-1}})_\pm = [m]_\pm \oplus \sum_{p=0}^{s-1} \oplus [2p], \tag{7.25}$$

$$(\mathbf{2^{m-1}})_\pm \otimes (\mathbf{2^{m-1}})_\mp = \sum_{p=0}^{s-1} \oplus [2p+1]; \tag{7.26}$$

<u>m odd</u>, say $m=2s+1$ so we are dealing with $so(4s+2)$:

$$(\mathbf{2^{m-1}})_\pm \otimes (\mathbf{2^{m-1}})_\mp = \sum_{p=0}^{s} \oplus [2p], \tag{7.27}$$

$$(\mathbf{2^{m-1}})_\pm \otimes (\mathbf{2^{m-1}})_\pm = [m]_\pm \oplus \sum_{p=0}^{s-1} \oplus [2p+1]. \tag{7.28}$$

In the above equations $[m]_\pm = P_\pm [m]$ and the tensor of rank m is treated separately because of its *duality* properties.

The duality concept arises as follows. We define for orthogonal transformations in n dimensions a rank n totally antisymmetric tensor ε by

$$\varepsilon_{\underbrace{ab\ldots}_{n}} = \begin{cases} 0 & \text{if any two indices are the same} \\ +1 & \text{if } ab\ldots \text{ is an even permutation of } 12\ldots, \\ -1 & \text{if } ab\ldots \text{ is an odd permutation of } 12\ldots \end{cases} \tag{7.29}$$

where we state these values in some Cartesian n-dimensional coordinate system, and the values in another system are to be obtained by transforming ε as a rank n tensor:

$$\varepsilon'_{\underbrace{AB\ldots K}_{n}} = \underbrace{O_{Aa} O_{Bb} \ldots O_{Kk}}_{n} \underbrace{\varepsilon_{ab\ldots k}}_{n}, \tag{7.30}$$

where O_{Aa}, etc., is the $n \times n$ orthogonal matrix for the transformation in question. It follows from the definition of the determinant that the expression on the right-hand side of (7.30) equals

$$\delta^1_{[A} \delta^2_{B} \ldots \delta^n_{K]} \det O, \tag{7.31}$$

where we have used for $\varepsilon_{ab\ldots}$ the values given by (7.29). But for $SO(n)$ we have $\det O = +1$ and so in fact $\varepsilon' = \varepsilon$, i.e. the ε tensor is an **invariant tensor**.

If we let $T_{\underbrace{ab...}_{A}}$ stand for the components of an antisymmetric tensor of rank A then we can associate with it an antisymmetric tensor of rank $n-A$ by forming the object

$$Q_{\underbrace{rt...}_{n-A}} = \varepsilon_{\underbrace{rt...\,ab...}_{n-A\ \ A}} T_{\underbrace{ab...}_{A}} / A! \qquad (7.32)$$

and we refer to the tensors Q and T as each others' **dual**.

Now then, when $A=n/2$, which of course means that n is even, $n=2m$, we have the situation where a tensor and its dual are of the same rank and we can therefore form tensors that are **self-dual** or **anti-self-dual**:

$$T+Q \quad \text{or} \quad T-Q. \qquad (7.33)$$

Thus, in $so(2m)$ the antisymmetric tensor of rank m is *not irreducible* and can be reduced into its self-dual and anti-self-dual parts.

To see what happens in detail consider the component $\gamma_1 \gamma_2 ... \gamma_m$ of the tensor of rank m. We have

$$\begin{aligned}
P_{\pm}\gamma_1\gamma_2...\gamma_m &= P_{\pm}(\pm\gamma_{2m+1})\gamma_1\gamma_2...\gamma_m \\
&= \pm(-i)^m P_{\pm}\gamma_1\gamma_2...\gamma_{2m}\gamma_1\gamma_2...\gamma_m \\
&= \pm(-i)^m(-1)^{m(3m-1)/2} P_{\pm}\gamma_{m+1}\gamma_{m+2}...\gamma_{2m} \qquad (7.34)\\
&= \pm(-i)^m(-1)^{m(3m-1)/2} \varepsilon_{\underbrace{12...\ ab...p}_{m\ \ \ \ m}} P_{\pm}\gamma_a\gamma_b...\gamma_p / m!.
\end{aligned}$$

Since

$$(-i)^m(-1)^{m(3m-1)/2} = \begin{cases} 1 \text{ for } m \text{ even} \\ i \text{ for } m \text{ odd} \end{cases} \qquad (7.35)$$

we conclude that $[\mathbf{2s}]_+$ is self-dual, $[\mathbf{2s}]_-$ is anti-self-dual, while $[\mathbf{2s+1}]_+$ and $[\mathbf{2s-1}]_-$ are each others' complex conjugate and appropriate (complex) linear combinations are self-dual and anti-self-dual. We further note that the dimensions of the reduced tensors are, of course, half as large as would be given by (7.18), i.e.

$$\dim[\mathbf{m}]_\pm = \tfrac{1}{2}\binom{2m}{m}. \qquad (7.36)$$

For the first few values of n these general results are displayed in detail in Table 7.1, where all representations are listed by their dimensions.

As is well known, among the n^2 matrices that span the space of $n\times n$ matrices $n(n+1)/2$ are symmetric and $n(n-1)/2$ are antisymmetric. This should manifest itself in application to our Clebsch–Gordan series in the behavior under transposition of the matrices $\gamma_A C^{-1}$, $A=0,1,2,\ldots,m$.

We have from (6.20) that

$$C^{\mathrm{T}}=(-1)^{m(m+1)/2}C \tag{7.37}$$

and from (6.23) that

$$C\gamma_j C^{-1}=(-1)^m \gamma_j^{\mathrm{T}}, \quad j=1,2,\ldots,2m,2m+1. \tag{7.38}$$

Therefore,

$$\underbrace{(\gamma_j \gamma_k \ldots \gamma_p)^{\mathrm{T}}}_{A} = \gamma_p^{\mathrm{T}} \ldots \gamma_k^{\mathrm{T}} \gamma_j^{\mathrm{T}} = (-1)^{Am} C\gamma_p \ldots \gamma_k \gamma_j C^{-1}$$

$$=(-1)^{Am+A(A-1)/2} C\gamma_j \gamma_k \ldots \gamma_p C^{-1}, \quad j<k<\ldots<p \tag{7.39}$$

Table 7.1 Glebsch–Gordan Series

Binomial series	$so(n)$	semispinor	spinor	Clebsch–Gordan series
1+2	$so(2)$	$1_+=\overline{1_-}$		$1_\pm \otimes 1_\mp = 1$, $1_\pm \otimes 1_\pm = \frac{2_\pm}{2}$
1+3	$so(3)$		$2^1=2$	$2\otimes 2=1\oplus 3$
1+4+6	$so(4)$	$2_\pm$		$2_\pm \otimes 2_\pm = 1\oplus \frac{6_\pm}{2}$, $2_\pm \otimes 2_\mp = 4$
1+5+10	$so(5)$		$2^2=4$	$4\otimes 4=1\oplus 5\oplus 10$
1+6+15+20	$so(6)$	$4_+=\overline{4_-}$		$4_\pm \otimes 4_\mp = 1\oplus 15$, $4_\pm \otimes 4_\pm = 6\oplus \frac{20_\pm}{2}$
1+7+21+35	$so(7)$		$2^3=8$	$8\otimes 8=1\oplus 7\oplus 21\oplus 35$
1+8+28+56+70	$so(8)$	$8_\pm$		$8_\pm \otimes 8_\pm = 1\oplus 28\oplus \frac{70_\pm}{2}$, $8_\pm \otimes 8_\mp = 8\oplus 56$
1+9+36+84+126	$so(9)$		$2^4=16$	$16\otimes 16=1\oplus 9\oplus 36\oplus 84\oplus 126$

so that

$$(\gamma_A C^{-1})^{\mathrm{T}} = C^{-1\mathrm{T}} \gamma_A^{\mathrm{T}} = (-1)^{m(m+1)/2} C^{-1} (-1)^{Am+A(A-1)/2} C \gamma_A C^{-1}$$
$$= (-1)^{(A-m)(A-m-1)/2} \gamma_A C^{-1}. \tag{7.40}$$

Thus, we arrive at the conclusion that the tensors of rank $A=m$ (mod 4) or $m-3$ (mod 4) are symmetric, while those of rank $A=m-1$ (mod 4) or $m-2$ (mod 4) are antisymmetric.

This means, for example, that the entries in the last line of Table 7.1 can also be expressed as

$$(\mathbf{16} \otimes \mathbf{16})_{\mathrm{sym}} = \mathbf{1} \oplus \mathbf{9} \oplus \mathbf{126}, \quad (\mathbf{16} \otimes \mathbf{16})_{\mathrm{antisym}} = \mathbf{36} \oplus \mathbf{84}$$

with the dimensions adding up to $16 \times 17/2$ and $16 \times 15/2$, respectively.

It also means that the trivial representation $A=0$ appears in the symmetric reduction for $m=0$ or 3 (mod 4) and in the antisymmetric reduction for $m=1$ or 2 (mod 4). In view of the reality properties of the spinors this agrees with the following statement, valid for any group: the trivial representation occurs in the reduction of $(S \otimes S)_{\mathrm{sym}}$ if S is orthogonal, in $(S \otimes S)_{\mathrm{antisym}}$ if S is symplectic and in $S \otimes \bar{S}$ if S is complex.

To conclude this chapter we comment on two situations in which duality plays an important role, the first occurs in $so(2)$ and the second in $so(4)$.

In $so(2)$ we have just one generator A_{12}, hence the group is Abelian and therefore all its irreducible representations are one-dimensional. Now clearly the adjoint is one-dimensional: $2(2-1)=1$ and the semispinors are one-dimensional: $2^{1-1}=1$, but the vector or defining representation is *two*-dimensional! Here is where duality comes to the rescue: in $so(2)$ the tensor whose dual is of the same rank is the vector—therefore the vector is reducible into the self-dual and anti-self-dual parts, each of dimension one.

In $so(4)$ the tensor whose dual if of the same rank is the antisymmetric tensor of rank two. Now we note that these are precisely the transformation properties of the generators of $so(n)$—therefore in $so(4)$ the six generators $A_{jk}=-A_{kj}$, $j,k=1,2,3,4$ are not irreducible and can be formed into the self-dual and anti-self-dual parts

$$A_{jk}^{\pm} = \tfrac{1}{2}(A_{jk} \pm \tfrac{1}{2}\varepsilon_{jksp} A_{sp}). \tag{7.41}$$

Quite explicitly we have

$$A^{\pm}_{12} = \tfrac{1}{2}(A_{12} \pm A_{34}),$$
$$A^{\pm}_{23} = \tfrac{1}{2}(A_{23} \pm A_{14}), \qquad (7.42)$$
$$A^{\pm}_{31} = \tfrac{1}{2}(A_{31} \pm A_{24}),$$

and the A^+ generators commute with the A^- generators. The result then is the reduction of $so(4)$ into two commuting parts. Since the three $A^+_{jk} = -A^+_{kj}$, $j,k=1,2,3$ generate $so(3)$ and so do the A^-_{jk} we can write the isomorphism

$$so(4) \cong so(3) \oplus so(3). \qquad (7.43)$$

We recognize the result that we found previously in Chapter 5—the proof offered here is perhaps more general since we do not require the realization of the generators via Clifford numbers.

8
The center and outer automorphisms of $Spin(n)$

Additional understanding of the different kinds of representations of $Spin(n)$ can be gained by looking at its center, that is at the elements that commute with all the elements of $Spin(n)$. Guided by the experience with $SO(3)$ and $SU(2) \cong Spin(3)$ we consider the element corresponding to rotation by the angle 2π in any plane. By definition, the element corresponding to rotation by the angle θ in the ij-plane is given by

$$R_{ij}(\theta) = \exp(i\theta A_{ij}). \tag{8.1}$$

It is straightforward to show that in the vector representation we have

$$(A_{ij}^2)_{sp} = \delta_{sp}(\delta_{si} + \delta_{sj}), \text{ (no sum)} \tag{8.2}$$

$$A_{ij}^3 = A_{ij} \tag{8.3}$$

and therefore $R_{ij}(\theta)$ is an $n \times n$ matrix whose only non-zero entries are $-\sin\theta$ at the intersection of the ith row and jth column, $\sin\theta$ at the transposed site, and ones on the diagonal except for $\cos\theta$ at the ith and the jth diagonal entry. Therefore

$$R_{ij}(2\pi) = \mathbf{1} \tag{8.4}$$

in the vector representation (and any representations obtained by tensoring the vector representation).

On the other hand, one readily shows that

$$\Gamma_{ij}^2 = (-\tfrac{i}{4}\gamma_{[i}\gamma_{j]})^2 = \tfrac{1}{4}\mathbf{1}, \tag{8.5}$$

so that

$$R_{ij}(\theta) = \mathbf{1}\cos\tfrac{\theta}{2} + 2i\Gamma_{ij}\sin\tfrac{\theta}{2} \tag{8.6}$$

and
$$R_{ij}(2\pi)=-1 \tag{8.7}$$
in the spinor representation.

Let us denote by E the element that stands for rotation by 2π in any plane. This element obviously commutes with every element of $Spin(n)$ but is not a constant (since it equals **1** in tensor, $-\mathbf{1}$ in spinor representations)—thus it belongs to the center of $Spin(n)$. The elements E and $E^2=\mathbf{1}$ form the two-element discrete group with the structure of $\mathbb{Z}_2$ and for $n=2m+1$ this is the end of the story but not so for $n=2m$.

For $n=2m$ we consider the element corresponding to simultaneous rotation by π in the m orthogonal planes $(1,2)$, $(3,4)$, ..., $(2m-1,2m)$. We denote this element by J and see that in the vector representation it is given by
$$J=-\mathbf{1}, \tag{8.8}$$
which is why it is called *inversion*. On the other hand, in the spinor representation it is given by
$$J=\exp i\pi(\Gamma_{12}+\Gamma_{34}+...+\Gamma_{2m-1,2m})$$
$$=(\exp\frac{\pi}{2}\gamma_1\gamma_2)(\exp\frac{\pi}{2}\gamma_3\gamma_4)...(\exp\frac{\pi}{2}\gamma_{2m-1}\gamma_{2m})$$
$$=\gamma_1\gamma_2\gamma_3\gamma_4...\gamma_{2m-1}\gamma_{2m}=i^m\gamma_{2m+1}, \tag{8.9}$$
which commutes with all the generators. Thus, the element J belongs to the center of $Spin(2m)$.

Since rotation by 2π in any plane equals E we have
$$J^2=E^m=\begin{cases}1 & \text{for } m \text{ even} \\ E & \text{for } m \text{ odd}\end{cases}. \tag{8.10}$$

That is to say the center of $Spin(2m)$ consists of the four elements
$$\mathbf{1},\ E,\ J,\ JE=EJ. \tag{8.11}$$

For even m these elements satisfy $J^2=\mathbf{1}$, $E^2=\mathbf{1}$, which is a discrete group isomorphic to $\mathbb{Z}_2\times\mathbb{Z}_2$, while for odd m they satisfy $J^2=E$, $J^3=EJ=JE$, $J^4=\mathbf{1}$, which is a discrete group isomorphic to $\mathbb{Z}_4$.

Since $\mathbb{Z}_k$ is isomorphic to the group whose elements are the kth roots of unity it follows that the four representations of the center of $Spin(2m)$ are for even m

J	E	JE	1
$+1$	$+1$	$+1$	$+1$
-1	$+1$	-1	$+1$
$+1$	-1	-1	$+1$
-1	-1	$+1$	$+1$

All these representations are self-conjugate and, as we know, all the representations of $Spin(4k)$ are self-conjugate (either orthogonal or symplectic). We remark on the curious fact that none of these representations are **faithful** (in a faithful representation different elements are represented differently).

For odd m the four representations of the center are obtained by representing J by the fourth roots of unity in turn:

J	$J^2{=}E$	$J^3{=}JE$	$J^4{=}1$
$+1$	$+1$	$+1$	$+1$
-1	$+1$	-1	$+1$
$+i$	-1	$-i$	$+1$
$-i$	-1	$+i$	$+1$

Two of these representations are self-conjugate and unfaithful and two are complex (and each others' conjugate) and faithful—it follows that faithful representations of $Spin(4k+2)$ are complex. In conclusion, we remark that the four realizations of J,E and JE correspond to the following four types of representations of $Spin(2m)$: the first row corresponds to even-rank tensors, the second row to odd-rank tensors and the last two rows correspond to the two semispinors.

Now what about the center of $SO(n)$? By definition, elements of $SO(n)$ are real orthogonal unimodular $n{\times}n$ matrices. The only matrix that commutes with all such matrices is a multiple of the unit matrix, $c\mathbf{1}$, and its determinant equals c^n. Since the only solutions to $c^n{=}{+}1$ over the reals are $c{=}{\pm}1$ for n even, $c{=}{+}1$ for n odd, it follows that the center of $SO(n)$ consists of the two elements $\pm\mathbf{1}$ (and is isomorphic to $\mathbb{Z}_2$) for n even, and consists of just $+\mathbf{1}$ and is trivial for n odd. Thus, we arrive at the isomorphism

$$Spin(n)/\mathbb{Z}_2 \cong SO(n), \tag{8.12}$$

where the $\mathbb{Z}_2$ consists of the elements $(E, \mathbf{1})$.

It is interesting that the $Spin(n)$ and $SO(n)$ groups exhibit these differences for even and odd n, even though the algebras look the same for even and odd n. We should like to offer an "explanation" of this feature by looking at automorphisms of $Spin(n)$. We recall that an isomorphism was defined as a one-to-one mapping between two algebraic structures that preserved all combinatorial operations associated with the structures. When the two structures are the same the isomorphism is called an **automorphism**.

In application to a Lie group G an automorphism is a one-to-one mapping:

$$g \to g', \quad g \in G, \quad g' \in G, \tag{8.13}$$

which preserves the group multiplication rule. If

$$g' = aga^{-1}, \quad a \in G, \tag{8.14}$$

then the automorphism is called **inner**. All other automorphisms are called **outer**. The inner automorphisms are simply similarity transformations, it is the outer automorphisms that are interesting.

In application to Lie algebras an automorphism is a mapping of the generators $X_a \to X'_a$ such that the structure constants are left unchanged:

$$[X_a, X_b] = if_{ab}{}^c X_c \Rightarrow [X'_a, X'_b] = if_{ab}{}^c X'_c. \tag{8.15}$$

Consider the conjugation

$$X_a \to X'_a = -X_a^T. \tag{8.16}$$

Clearly we have here an automorphism. If it is inner the representation is self-conjugate. Thus, we have the conclusion: complex representations can occur only if the algebra in question has *outer* automorphisms. This condition, while necessary, is not sufficient. We shall show later that a sufficient condition is the existence of independent Casimir operators of *odd* degree.

It is possible to classify all semisimple algebras in terms of so-called Dynkin diagrams, which encode all the information contained in the commutation relations and it will follow that outer automorphisms correspond to symmetries of the Dynkin diagram. In the following, in application to $Spin(n)$ we take a more direct approach and argue that outer automorphisms arise as a result of the existence of *reflections*.

Concentrating on $Spin(n)$ and its generators $A_{kj}=-A_{jk}$, $k,j=1,2,\ldots,n$, we consider the following mapping:

$$A_{kj} \to A'_{kj} = \begin{cases} -A_{kj} & \text{if } k=1 \text{ or } j=1 \\ +A_{kj} & \text{if } k\neq 1 \text{ and } j\neq 1 \end{cases}. \tag{8.17}$$

This is clearly an automorphism since the commutation relations are unaffected. To determine whether it is inner or outer we observe that in the defining representation the $O(n)$ transformation, which inverts axis one and leaves all other axes unchanged is the $n \times n$ diagonal matrix

$$a = \text{diag}(-1, 1, 1, \ldots, 1) \tag{8.18}$$

and the mapping (8.17) can be accomplished by $\pm a$. Since

$$\det a = -1, \quad \det(-a) = (-1)^{n-1} \tag{8.19}$$

for n odd $-a$ is a rotation, an element of $SO(n)$, and the mapping can be achieved as an inner automorphism, i.e. a triviality. But for n even, both a and $-a$ are reflections, not rotations, and we have an outer automorphism. We show next that the two semispinors are interchanged under this automorphism—they owe their existence to this automorphism.

To see that the semispinors are interchanged we note that the element γ_{2m+1}, which commuted with all the generators of $so(2m)$ and was used to project out semispinors, changes sign under the reflections $\pm a$ since it is a product of $2m$ γs, an odd number of which change sign. Lastly, we see that under the $O(2m)$ group, i.e. the $SO(2m)$ group plus reflections, the original 2^m-dimensional spinor is irreducible—the reduction into the two semispinors is valid only when reflections are *not* included.

We might inquire into the action of this automorphism on the elements of the center. Writing

$$E = \exp 2i\pi A_{12},$$
$$J = \exp i\pi(A_{12}+A_{34}+\ldots+A_{2m-1,2m}), \tag{8.20}$$
$$JE = \exp i\pi(-A_{12}+A_{34}+\ldots+A_{2m-1,2m}),$$

we see that the mapping (8.17) exchanges J and JE and leaves E unchanged. That makes one wonder what mapping, if any, would instead exchange, say, E and JE, and leave J unchanged? Thus, we want a mapping that contains in particular

$$A_{12} \to \tfrac{1}{2}(\pm A_{12} \pm A_{34} \pm ... \pm A_{2m-1,2m}) \qquad (8.21)$$

with some appropriate choice of signs. Because the length squared of the two sides of (8.21) differs by a factor $m/4$ this mapping cannot be an automorphism except possibly for $m=4$. Indeed, for $Spin(8)$ such an outer automorphism exists and has the effect of exchanging one of the semispinors with the vector (both being 8-dimensional).

We conclude this chapter by listing some isomorphisms involving $Spin(n)$ and its factor groups:

$$Spin(3) \cong SU(2),$$
$$Spin(3)/\mathbb{Z}_2 \cong SO(3)$$
$$Spin(4) \cong Spin(3) \otimes Spin(3)$$
$$Spin(4)/\mathbb{Z}_2 \cong SO(4)$$
$$Spin(4)/\mathbb{Z}_2 \times \mathbb{Z}_2 \cong SO(3) \otimes SO(3)$$
$$Spin(5) \cong Sp(4)$$
$$Spin(5)/\mathbb{Z}_2 \cong SO(5)$$
$$Spin(6) \cong SU(4)$$
$$Spin(6)/\mathbb{Z}_2 \cong SO(6).$$

Of these the ones involving the symplectic group $Sp(4)$ and the unitary group $SU(4)$ have not yet been demonstrated.

For $n>6$ the only isomorphisms to any classical groups are $Spin(n)/\mathbb{Z}_2 \cong SO(n)$.

Biographical Sketch

Dynkin, Eugene Borisovich (1924–) was born in Leningrad, USSR (now St. Petersburg, Russia). He received his Ph.D. from Moscow University in 1948 and was on the faculty there until 1968. He was a senior research scholar at the Academy of Sciences, USSR, from 1968 to 1976 and went to Cornell University as a professor of Mathematics in 1977. His name is attached to various concepts in group theory: Dynkin diagrams, Dynkin labels, Dynkin index, etc.

9
Composition algebras

In previous chapters Clifford numbers played an important role in the construction of spin representations of orthogonal algebras. Certain of the techniques developed in connection with Clifford numbers turn out to be useful in the proof of a theorem due to Hurwitz, which asserts that there are only four normed composition algebras with unit element—the real numbers $\mathbb{R}$, the complex numbers $\mathbb{C}$, the quaternions $\mathbb{H}$, and the octonions $\mathbb{O}$.

It was Hamilton (in whose honor quaternions are denoted by $\mathbb{H}$) who proposed to treat complex numbers as pairs of real numbers and discovered that a generalization of complex numbers was possible involving a quartet of real numbers—the quaternions. This was further generalized by Graves and, independently, by Cayley to an algebra involving an octet of real numbers—the octonions. The theorem of Hurwitz states that if certain nice properties are desired, such as absence of zero divisors, then these four—$\mathbb{R}$, $\mathbb{C}$, $\mathbb{H}$ and $\mathbb{O}$—are the only possibilities. We should also mention Frobenius' theorem, which states that every associative division algebra is isomorphic to $\mathbb{R}$, $\mathbb{C}$ or $\mathbb{H}$

As a warm-up consider a **complex** number X, its **conjugate** $\overline{X}$, and its **norm** squared $\|X\|$:

$$X = x_0 + ix_1, \quad \overline{X} = x_0 - ix_1, \quad \|X\| = X\overline{X} = x_0^2 + x_1^2, \qquad (9.1)$$
$$x_0, x_1 \in \mathbb{R}, \quad i^2 = -1.$$

It is easy to verify that the norm satisfies the composition law

$$\|XY\| = \|X\| \|Y\|. \qquad (9.2)$$

Suppose now that we have $N-1$ imaginary units e_k, $k=1,2,\ldots,N-1$, and consider

$$X = x_0 + x_k e_k, \quad \overline{X} = x_0 - x_k e_k, \quad x_0, x_k \in \mathbb{R}. \qquad (9.3)$$

Such Xs will form an algebra if we specify the multiplication rule for the e_k. In particular, if these imaginary units satisfy

$$e_k e_l + e_l e_k = -2\delta_{kl} \tag{9.4}$$

then

$$\|X\| = X\overline{X} = x_0^2 + x_k^2 \tag{9.5}$$

and we have a **division algebra** in the sense that every X, which is not identically zero, has an inverse given by

$$X^{-1} = \overline{X}/\|X\| \tag{9.6}$$

and we have a **composition algebra** if

$$\|XY\| = \|X\| \|Y\|. \tag{9.7}$$

So to obtain a composition algebra we need to find a multiplication rule for the e_k, which contains (9.4) and satisfies the constraint (9.7). It is convenient to define

$$e_0 = 1 \tag{9.8}$$

and state the multiplication rule as

$$e_\alpha e_\beta = e_\gamma h_{\gamma\alpha\beta}, \tag{9.9}$$

where Greek subscripts run over $0, 1, 2, \ldots, N-1$ and where the structure constants $h_{\gamma\alpha\beta}$ are real numbers. As a consequence of $e_0 e_\alpha = e_\alpha e_0 = e_\alpha$ we have

$$h_{\gamma 0 \alpha} = h_{\gamma \alpha 0} = \delta_{\gamma\alpha}, \tag{9.10}$$

and as a consequence of (9.4) we have

$$h_{a(bc)} = 0, \tag{9.11}$$

where the brackets ( ) denote symmetrizing.

Now, we can write

$$X = x_\alpha e_\alpha, \quad \|X\| = x_\alpha x_\alpha, \quad XY = x_\alpha y_\beta e_\gamma h_{\gamma\alpha\beta} \tag{9.12}$$

and the constraint (9.7) becomes

$$x_\alpha y_\beta h_{\gamma\alpha\beta} x_\mu y_\nu h_{\gamma\mu\nu} = x_\alpha x_\alpha y_\beta y_\beta, \tag{9.13}$$

which requires

$$h_{\gamma\alpha\beta}h_{\gamma\mu\nu}+h_{\gamma\mu\beta}h_{\gamma\alpha\nu}=2\delta_{\alpha\mu}\delta_{\beta\nu}. \tag{9.14}$$

It is convenient to view the $h_{...}$ as elements of $N\times N$ real matrices:

$$h_{\gamma\alpha\beta}=(h_\alpha)_{\gamma\beta} \tag{9.15}$$

in terms of which (9.14) becomes

$$h_\alpha^T h_\mu + h_\mu^T h_\alpha = 2\delta_{\alpha\mu}1_N, \tag{9.16}$$

where 1_N denotes the $N\times N$ unit matrix. Next, we observe that part of (9.10) becomes

$$h_0=1_N \tag{9.17}$$

and for $N=1$ this is the end of the story giving us the real numbers $\mathbb{R}$.

For $N\geq 2$ we take in (9.16) $\mu=0$ and learn that

$$h_a^T+h_a=0. \tag{9.18}$$

Since Latin subscripts run over $1,2,...,N-1$ we conclude that the $N-1$ matrices h_a must be real antisymmetric $N\times N$ matrices obeying

$$h_a h_b + h_b h_a = -2\delta_{ab}1_N, \tag{9.19}$$

$$(h_a)_{\nu 0}=\delta_{\nu a},\ h_{abc}\ \text{totally antisymmetric.} \tag{9.20}$$

Clearly (9.19) will be satisfied if we take

$$h_a=-i\gamma_a \tag{9.21}$$

with γ_a the previously discussed Clifford numbers. In other words, the ih_a are a special case of the Clifford numbers that are pure imaginary antisymmetric and satisfy (9.20). Since the h_a are $N\times N$ and Cliffords are $2^m\times 2^m$ we must have

$$N=2^m \tag{9.22}$$

and since there are $2m+1$ of the Cliffords and $N-1$ of the h_a we must have $2m+1\geq N-1$ or

$$2m+2\geq 2^m. \tag{9.23}$$

The left-hand side of (9.23) grows linearly with m while the right-hand side grows exponentially. Thus, the inequality must fail for some sufficiently large m and in fact we have by inspection that only $m=1,2$ and 3 are allowed (The $N=1$ case can be included by including $m=0$).

The conclusion that m must be 1, 2, or 3 is a necessary requirement—we must still check that the correct number of pure imaginary antisymmetric Clifford matrices to serve as ih_a exist and that (9.20) can be satisfied.

m=1, N=2. We need one real antisymmetric 2×2 matrix, which of course is available:

$$h_1 = -i\sigma = \begin{pmatrix} 0 & -1 \\ 1 & 0 \end{pmatrix}. \tag{9.24}$$

With this choice for h_1 and the rows and columns labeled 0 and 1 we have $(h_1)_{m0} = \delta_{m1}$, i.e. (9.20) is satisfied and we are finished. We recognize that we have complex numbers $\mathbb{C}$ since $e_0 = 1$ and e_1 behaves like i because

$$e_1^2 = e_\alpha h_{\alpha 11} = e_0(h_1)_{01} + e_1(h_1)_{11} = -1. \tag{9.25}$$

Moreover, since the multiplication of 1 and i is commutative and associative we recover the fact that $\mathbb{C}$ constitutes a field.

Lastly, we remark that we have found here a representation of complex numbers by 2×2 *real* matrices:

$$1 \to 1_2 = \begin{pmatrix} 1 & 0 \\ 0 & 1 \end{pmatrix}, \quad i \to h_1 = \begin{pmatrix} 0 & -1 \\ 1 & 0 \end{pmatrix}. \tag{9.26}$$

m=2, N=4. We need three real antisymmetric 4×4 matrices. Since $i\sigma$ is real antisymmetric, ρ and τ are real symmetric, the following are satisfactory:

$$i\sigma \otimes 1_2, \quad \rho \otimes i\sigma, \quad \tau \otimes i\sigma \tag{9.27}$$

and in fact there is also another choice

$$1_2 \otimes i\sigma, \quad i\sigma \otimes \rho, \quad i\sigma \otimes \tau \tag{9.28}$$

and every matrix from the first set commutes with every matrix from the second set. Note also that the total number of antisymmetric 4×4 matrices is six and we have found all of them. This "six" corresponds to the fact that we are dealing here with so(4) (by definition the algebra generated by 4×4 real antisymmetric matrices) that has six generators, and the break-up into the two commuting sets is a manifestation of the isomorphism $so(4) \cong so(3) \oplus so(3)$.

Explicitly

$$h_1 = -1_2 \otimes i\sigma = \begin{pmatrix} -i\sigma & 0_2 \\ 0_2 & -i\sigma \end{pmatrix} = \begin{pmatrix} 0 & -1 & 0 & 0 \\ 1 & 0 & 0 & 0 \\ 0 & 0 & 0 & -1 \\ 0 & 0 & 1 & 0 \end{pmatrix} \qquad (9.29)$$

$$h_2 = -i\sigma \otimes \tau = \begin{pmatrix} 0_2 & -\tau \\ \tau & 0_2 \end{pmatrix} = \begin{pmatrix} 0 & 0 & -1 & 0 \\ 0 & 0 & 0 & 1 \\ 1 & 0 & 0 & 0 \\ 0 & -1 & 0 & 0 \end{pmatrix} \qquad (9.30)$$

$$h_3 = -i\sigma \otimes \rho = \begin{pmatrix} 0_2 & -\rho \\ \rho & 0_2 \end{pmatrix} = \begin{pmatrix} 0 & 0 & 0 & -1 \\ 0 & 0 & -1 & 0 \\ 0 & 1 & 0 & 0 \\ 1 & 0 & 0 & 0 \end{pmatrix} \qquad (9.31)$$

are a choice that satisfies (9.20) when the rows and columns are labeled 0, 1, 2, 3. Moreover, it is seen by inspection that

$$(h_b)_{ac} = \varepsilon_{abc}, \qquad (9.32)$$

which combined with $h_{0ba} = -h_{ab0} = -\delta_{ab}$ [which follows from (9.20) and antisymmetry of h_b] leads to

$$e_a e_b = -\delta_{ab} + e_c \varepsilon_{abc}. \qquad (9.33)$$

Equation (9.33) is the standard definition of **quaternions** ℍ. We note that multiplication of quaternions is no longer commutative but still associative—thus the quaternions ℍ are a non-commutative field.

We note further that

$$(h_a h_b)^{\mathrm{T}} = h_b^{\mathrm{T}} h_a^{\mathrm{T}} = h_b h_a = -h_a h_b, \qquad (9.35)$$

so the three hs defined by (9.29), (9.30) and (9.31) could close under ordinary matrix multiplication. Indeed explicit calculation shows that

$$h_a h_b = -\delta_{ab} 1_4 + h_c \varepsilon_{abc} \qquad (9.36)$$

and therefore we have a representation of quaternions by *real* 4×4 matrices:

$$1 \to 1_4, \quad e_a \to h_a, \quad a=1,2,3. \qquad (9.37)$$

In view of the occurrence of the epsilon symbol, which is the structure constant of $su(2)$, it is perhaps not surprising that we can also represent quaternions by 2×2 *complex* matrices:

$$1 \to 1_2 = \begin{pmatrix} 1 & 0 \\ 0 & 1 \end{pmatrix}, \quad e_1 \to i\sigma = \begin{pmatrix} 0 & 1 \\ -1 & 0 \end{pmatrix},$$

$$e_2 \to i\rho = \begin{pmatrix} 0 & i \\ i & 0 \end{pmatrix}, \quad e_3 \to i\tau = \begin{pmatrix} i & 0 \\ 0 & -i \end{pmatrix}. \qquad (9.38)$$

$m=3, N=8$. Here we need *seven* real antisymmetric 8×8 matrices. With the symmetry and reality properties of our basic 2×2 matrices in mind we see that the following choice satisfies all requirements [including (9.20), if rows and columns are labeled 0, 1, 2,..., 7]:

$$h_1 = \tau \otimes 1_2 \otimes (-i\sigma)$$
$$h_2 = 1_2 \otimes (-i\sigma) \otimes \tau$$
$$h_3 - 1_2 \otimes (-i\sigma) \otimes \rho$$
$$h_4 = -i\sigma \otimes \tau \otimes 1_2 \qquad (9.39)$$
$$h_5 = \rho \otimes 1_2 \otimes (-i\sigma)$$
$$h_6 = -i\sigma \otimes \rho \otimes 1_2$$
$$h_7 = i\sigma \otimes \sigma \otimes \sigma.$$

Explicitly

$$h_1 = \tau \otimes 1_2 \otimes (-i\sigma) = \begin{pmatrix} 1_2 \otimes (-i\sigma) & 0_4 \\ 0_4 & 1_2 \otimes i\sigma \end{pmatrix} = \begin{pmatrix} -i\sigma & 0_2 & 0_2 & 0_2 \\ 0_2 & -i\sigma & 0_2 & 0_2 \\ 0_2 & 0_2 & i\sigma & 0_2 \\ 0_2 & 0_2 & 0_2 & i\sigma \end{pmatrix}$$

$$= \begin{pmatrix} 0 & -1 & 0 & 0 & 0 & 0 & 0 & 0 \\ 1 & 0 & 0 & 0 & 0 & 0 & 0 & 0 \\ 0 & 0 & 0 & -1 & 0 & 0 & 0 & 0 \\ 0 & 0 & 1 & 0 & 0 & 0 & 0 & 0 \\ 0 & 0 & 0 & 0 & 0 & 1 & 0 & 0 \\ 0 & 0 & 0 & 0 & -1 & 0 & 0 & 0 \\ 0 & 0 & 0 & 0 & 0 & 0 & 0 & 1 \\ 0 & 0 & 0 & 0 & 0 & 0 & -1 & 0 \end{pmatrix}, \qquad (9.40)$$

etc. We may summarize the results by the matrix below from which we obtain the matrix h_k by putting 1 in the entry marked k, -1 in the entry marked $-k$ and zeros everywhere else:

$$\begin{pmatrix} 0 & -1 & -2 & -3 & -4 & -5 & -6 & -7 \\ 1 & 0 & -3 & 2 & 5 & -4 & 7 & -6 \\ 2 & 3 & 0 & -1 & -6 & 7 & 4 & -5 \\ 3 & -2 & 1 & 0 & -7 & -6 & 5 & 4 \\ 4 & -5 & 6 & 7 & 0 & 1 & -2 & -3 \\ 5 & 4 & -7 & 6 & -1 & 0 & -3 & 2 \\ 6 & -7 & -4 & -5 & 2 & 3 & 0 & 1 \\ 7 & 6 & 5 & -4 & 3 & -2 & -1 & 0 \end{pmatrix}. \qquad (9.41)$$

We may use these h_k to write out (9.9) for octonions in the form

$$e_a e_b = -\delta_{ab} + h_{abc} e_c. \qquad (9.42)$$

It follows from (9.41) that h_{abc}, which is completely antisymmetric, is equal to $+1$ for

$$abc = 123,\ 154,\ 176,\ 246,\ 275,\ 347,\ 356 \qquad (9.43)$$

(we note that this is but one of 480 distinct ways of labeling the octonions).

Octonion multiplication is obviously non-commutative. It is also, in agreement with the Frobenius' theorem, *non-associative*. To see this consider the equation

$$e_1 e_2 = e_3 \qquad (9.44)$$

and, assuming associativity, commute e_k across both sides. For $k \neq 1,2,3$ e_k commutes with the left side and anticommutes with the right side, since it anticommutes with each of the e_1, e_2 and e_3. Consequently, if there are more than three of the anticommuting imaginary units e_k the resultant algebra *must* be non-associative. In particular, this means that octonions cannot be represented by matrices since matrix multiplication is associative.

The **associator** $[x,y,z]$ of three octonions x, y, z is defined by

$$[x,y,z] \equiv (xy)z - x(yz) \qquad (9.45)$$

and is a measure of the lack of associativity, similarly to the way the commutator $[a,b]$ measures the lack of commutativity. We may introduce a four-indexed object h_{abcd} related to the associator by

$$[e_a, e_b, e_c] = 2 h_{abcd} e_d. \qquad (9.46)$$

It follows from the definition of the associator and from (9.42) that

$$2 h_{abcd} = \delta_{b[c} \delta_{a]d} + h_{bk[c} h_{a]kd}, \qquad (9.47)$$

which is obviously antisymmetric in a and c (and b and d). In fact h_{abcd} is completely antisymmetric. To see that, consider symmetrizing in, say, a and b:

$$\begin{aligned} 2 h_{(ab)cd} &= \delta_{b[c} \delta_{a]d} + h_{bk[c} h_{a]kd} + \delta_{a[c} \delta_{b]d} + h_{ak[c} h_{b]kd} \\ &= \delta_{c[b} \delta_{a]d} - 2 \delta_{ab} \delta_{cd} - h_{cak} h_{kbd} - h_{cbk} h_{kad} \\ &= -h_{ca0} h_{0bd} - h_{cb0} h_{0ad} - 2 \delta_{ab} \delta_{cd} - h_{cak} h_{kbd} - h_{cbk} h_{kad} \\ &= -(2 \delta_{ab} 1_8 + h_a h_b + h_b h_a)_{cd} = 0, \end{aligned} \qquad (9.48)$$

where we have used (9.18–9.20).

For the associator to be non-zero not only must the three e_a, e_b, e_c be all different, we must also have $e_a \neq \pm e_b e_c$ in order not to be dealing with a subalgebra of the octonions corresponding to quaternions. As we know, quaternions are associative and one can verify that (9.47) gives zero for h_{abcd} when the three-indexed hs are taken equal to the three-indexed epsilons as is appropriate for quaternions.

Calculating explicitly the values of h_{abcd} we find it equal to the $so(7)$ dual of h_{abc}, meaning

$$h_{abcd} = \varepsilon_{abcdefg} h_{efg}/3! \qquad (9.49)$$

In the next chapter we shall make use of these objects to discuss the exceptional group G_2.

Before leaving composition algebras we mention a curiosity called the two, four and eight squares theorems. Since the norm squared is, respectively, the sum of two, four and eight real squares for $\mathbb{C}$, $\mathbb{H}$ and $\mathbb{O}$ the composition law (9.7) states that the product of a sum of, respectively, two, four or eight squares times the sum of, respectively, two, four or eight squares equals the sum of, respectively, two, four or eight squares. This has been studied in number theory as it applies to sums of squares of integers: the two-square version was known to Diophantus, the four-square version was shown by Euler and the eight-square version is usually credited to Graves, the discoverer of octonions, although it was actually published some 20 years earlier by the Danish mathematician Degen. Degen believed that 2^m versions should exist for any m, but as we know from Hurwitz only $m=0, 1, 2$ and 3 are possible.

Biographical Sketches

Hurwitz, Adolph (1859–1919) was born in Hildesheim, Germany. In 1884 he became professor at Köningsberg University where Hilbert and Minkowski were among his students. In 1892 he accepted Frobenius' chair at the Zürich Polytechnic University. Hurwitz's theorem has been extended several times with the final result by Milnor that algebras over the reals without zero divisors exist in 1, 2, 4 and 8 dimensions only. He died in Zürich.

Hamilton, Sir William Rowan (1805–65) was born in Dublin, Ireland. He was a child prodigy, who was appointed Andrews Professor of Astronomy at Trinity College and Astronomer Royal for Ireland at the age of 22. He made major contributions to dynamics with Hamilton's equations (= equations of motion) involving the Hamiltonian. Hamilton's principle asserts that the action integral is an extremum. The Hamilton–Jacobi formalism is viewed by many as the precursor of wave mechanics. His discovery of quaternions followed after many fruitless years of trying to find "triplets" that would answer the question: is it possible to find a hypercomplex number that is related to 3-dimensional space just as ordinary complex numbers are related to 2-dimensional space? Towards the end of his life he drank increasingly, eventually dying of gout at Dunsink Observatory near Dublin.

Graves, John Thomas (1806–70) was born in Dublin, Ireland. He was a barrister, named professor of jurisprudence at University College, London. He corresponded a great deal with Hamilton. Within two months of Hamilton's discovery of quaternions Graves found that hypercomplex numbers composed of eight elements, which he called "octaves", also satisfied the composition law and he described them in a letter to Hamilton in December 1843. Hamilton promised to make it public but delayed for various reasons and Cayley published his results in March 1845 and octonions became known as Cayley numbers.

Cayley, Arthur (1821–95) was born in Richmond, England. Unwilling to take holy orders—at that time a requirement for a mathematical career at Cambridge—he was forced to spend 14 years as a barrister, during which time he wrote 300 mathematical papers. When the requirement was dropped he returned to Cambridge becoming in 1863 the Sadlerian Professor there. He was a prolific mathematician (collected works fill 13 volumes), who created the theory of invariants with his friend J. J. Sylvester. His name is attached to many concepts: the Cayley–Hamilton theorem stating that a matrix satisfies its own characteristic equation; the Cayley–Klein parameters providing a description of rotations; the Cayley numbers, i.e. the octonions. Perhaps the best known of the many Cayley theorems is that every finite group whatsoever is isomorphic to a suitable group of permutations. One of Cayley's notable non-mathematical achievements was his championing the admission of women students to the University of Cambridge. He died in Cambridge.

Frobenius, Ferdinand Georg (1849–1917) was born in Berlin, Germany. He worked on the theory of finite groups and group characters. He taught at Zürich from 1875 to 1892 when he left to become professor at the University of Berlin. He collaborated with Schur, with whom he developed the representation theory of finite groups through linear substitutions.

10
The exceptional group G_2

As was shown in the preceding chapter the commutator and associator of octonions provide us with two antisymmetric entities h_{abc} and h_{abcd} related by

$$2h_{abcd} = \delta_{b[ca]d} + h_{bk[c}h_{a]kd}, \tag{10.1}$$

where we have introduced the abbreviation

$$\delta_{abcd} = \delta_{ab}\delta_{cd}. \tag{10.2}$$

Products of these hs satisfy a number of identities. The following identity

$$h_{amn}h_{bmn} = 6\,\delta_{ab}, \tag{10.3}$$

is proved by noting that for a given m and n h_{amn} is non-zero for a unique value of a, therefore $h_{amn}h_{bmn}$ must be proportional to δ_{ab}. The constant of proportionality is determined to be 6 since for a given a there are 6 choices for $m \neq a$, and for a given a and m there is according to (9.43) a unique n such that h_{amn} is non-zero.

Next, we prove

$$h_{amn}h_{bnp}h_{cmp} = 3\,h_{abc} \text{ (equivalently } h_{amn}h_{bcmn} = 4\,h_{abc}\text{)}. \tag{10.4}$$

We observe by inspection that $h_{amn}h_{bnp}h_{cmp}$ is completely antisymmetric in a,b and c, hence proportional to h_{abc}. To determine the proportionality constant we observe that given a and b there is a unique c such that h_{abc} is non-zero. Now, there are 5 choices for m not equal to a or c, and for a given a, b, c and m there are unique choices for n and p such that the various hs are non-zero. According to (9.43) four of the choices for m result in $+h_{abc}$, while the fifth choice results in $-h_{abc}$. This completes the proof of (10.4).

We list several more identities:

$$h_{mc(a}h_{b)dm} = \delta_{c(ab)d} - \delta_{(ab)cd}, \tag{10.5}$$

$$h_{mc[a}h_{b]dm} = 2\, h_{abn}h_{cdn} + 3\delta_{d[ab]c}, \tag{10.6}$$

$$2h_{abcm}h_{dem} = \delta^{[e}{}_{[a}h_{bc]}{}^{d]}, \tag{10.7}$$

$$h_{ab(c}h_{d)ef} = h^{[a}{}_{f(c}h_{d)e}{}^{b]} + \delta^{[a}{}_{(c}h^{b]}{}_{d)ef} + \delta^{[e}{}_{(c}h^{f]}{}_{d)ab} + \delta_{(cd)}h_{abef}, \tag{10.8}$$

$$2h_{ab[c}h_{d]ef} = h_{cd[e}h_{f]ab} - h_{cd[a}h_{b]ef} + \delta^{[c}{}_{[e}h_{f]ab}{}^{d]}$$
$$- \delta^{[c}{}_{[a}h_{b]ef}{}^{d]} - 2\delta^{[e}{}_{[a}h_{b]cd}{}^{f]} \tag{10.9}$$

$$12h_{abcm}h_{defm} = 12\delta_{[a}{}^{d}{}_{b}{}^{e}{}_{c]}{}^{f} + \delta^{[d}{}_{[a}h^{ef]}{}_{bc]} - h^{[d}{}_{[ab}h^{ef]}{}_{c]}. \tag{10.10}$$

Equations (10.5) – (10.10) are proved by writing down on the right all possible terms involving δs and hs with the indices appropriately symmetrized to correspond to the symmetries on the left. The coefficients of these terms are then determined by contraction with a δ or an h, thus reducing it to an earlier equation.

We can use (10.6) to rewrite (10.1) in the following more useful form:

$$h_{abm}h_{cdm} = h_{abcd} + \delta_{c[ab]d}. \tag{10.11}$$

We now use these octonionic structure constants to introduce the exceptional group G_2 as a subgroup of $SO(7)$. We do this by considering the following two subsets (denoted by G_{ab} and G_a) of the generators A_{ab} of $so(7)$:

$$G_{ab} = 2\, A_{ab} - \tfrac{1}{2} h_{abmn} A_{mn}, \tag{10.12}$$

$$G_a = h_{amn} A_{mn}. \tag{10.13}$$

It follows from these definitions that

$$h_{abm} G_m = h_{abm} h_{mpq} A_{pq} = (h_{abpq} + \delta_{a[p|b]}) A_{pq} = -2G_{ab} + 6A_{ab}$$

i.e. we have the decomposition of the 21 generators of $so(7)$ as

$$6A_{ab} = 2G_{ab} + h_{abm} G_m. \tag{10.14}$$

While it is clear that there are seven of the G_a it would seem that since $G_{ab} = -G_{ba}$ this defines 21 entities, however, as a consequence of (10.4) we have

$$h_{abc}G_{ab}=0, \qquad (10.15)$$

which are seven constraints so that only 14 of the G_{ab} are linearly independent.

In fact, the subset G_{ab} generates a subalgebra, the so-called g_2 algebra (and the corresponding exceptional group G_2), and the subset G_a transforms as the seven-dimensional representation of g_2. To see this we form the commutators of G_{ab} with themselves and with G_a and find, using the identities (10.3) – (10.11) and the $so(7)$ commutation relations for the A_{mn} as needed, that

$$[G_{ab}, G_m] = i(2\delta_{m[a}G_{b]} - h_{mabk}G_k), \qquad (10.16)$$

$$[G_{ab}, G_{mn}] = i(2\delta_{m[a}G_{b]n} - h_{mabk}G_{kn} - 2G_{m[a}\delta_{b]n} - h_{nabk}G_{mk}). \qquad (10.17)$$

We also conclude that (10.14) can be viewed as showing that the **21** adjoint representation of $so(7)$, while irreducible under $so(7)$, reduces under restriction to g_2 as

$$\mathbf{21 = 14 \oplus 7}. \qquad (10.18)$$

In Chapter 17 we shall give a different demonstration that G_2 is a subgroup of $SO(7)$ by looking at the root spaces of these groups.

Next we show that the exceptional group G_2 is the group of automorphisms of octonions $\mathbb{O}$. Consider replacing the seven octonions e_a by some linear combinations

$$e'_a = C_{ab}e_b, \qquad (10.19)$$

where the primed octonions obey precisely the same multiplication rules as the unprimed ones, i.e. with the same structure constants h_{abc}. That means that the C_{ab} are elements of a group of transformations under which h_{abc} transforms as an *invariant tensor*:

$$h'_{abc} = C_{am}C_{bn}C_{ck}h_{mnk} = h_{abc}. \qquad (10.20)$$

Equivalently, if G_A are the infinitesimal generators of this group of transformations then we can rewrite (10.20) as

$$0 = h'_{abc} - h_{abc} = (G_A)_{am}h_{mbc} + (G_A)_{bm}h_{amc} + (G_A)_{cm}h_{abm}. \qquad (10.21)$$

Now consider the generators of G_2 as defined by (10.12) and insert for A_{mn} the $so(7)$ generators in the defining seven-dimensional representation:

$$(G_{ab})_{mn} = i(h_{abmn} - 2\delta_{a[m n]b}). \tag{10.22}$$

It then follows that

$$2(h'_{abc} - h_{abc}) = -(G_{pq})_{m[a} h_{bc]m} = -i(h_{pqm[a} - 2\delta_{m[pq][a}) h_{bc]m}$$
$$= -i(h_{pqm[a} h_{bc]m} + 2\,\delta^{[p}_{[a} h_{bc]}{}^{q]}) = 0 \tag{10.23}$$

as a consequence of (10.7). This proves that h_{abc} is invariant under G_2 transformations, i.e. that G_2 is the automorphism group of the octonions $\mathbb{O}$.

While we are on this subject we could ask what is the automorphism group of quaternions $\mathbb{H}$ and the complex numbers $\mathbb{C}$. If we recall the definition of quaternions by (9.33)

$$e_a e_b = -\delta_{ab} + e_c \varepsilon_{abc}, \qquad a,b,c = 1,2,3$$

then we see that the group of autmorphisms of $\mathbb{H}$ must leave invariant ε_{abc}, which, of course, is $SO(3)$.

For the complex numbers $\mathbb{C}$ the only manipulation of the sole imaginary unit i that leaves the multiplication table unchanged is to change it to -i, so the automorphism group of $\mathbb{C}$ is $\mathbb{Z}_2 = \{1, -1\}$.

The G_2 group was introduced into Physics by Racah in his groundbreaking papers on the theory of atomic spectra. In considering equivalent electrons in an atom in the shell specified by the angular momentum l Racah found it necessary to consider the chain of groups

$$SU(2l+1) \supset SO(2l+1) \supset SO(3). \tag{10.24}$$

In restricting a group to a subgroup, irreducible representations of the group decompose into a sum of irreducible representations of the subgroup. The decomposition of a given representation of the group has the desirable property of being *multiplicity-free* if representations of the subgroup occur either once or not at all.

Racah showed that a basis for the configuration l^n for n equivalent electrons in the l shell, $0 \leq n \leq 2l+1$, was provided by representations of $SU(2l+1)$. These can be specified by the partition $(\lambda_1, \lambda_2, \ldots, \lambda_{2l+1})$ (see Chapters 14 and 15). Because of the Pauli exclusion principle only a limited number of representations of $SU(2l+1)$ are allowed, namely those for which $0 \leq \lambda_i \leq 2$. For the configurations corresponding to the s^n, p^n and d^n electrons (i.e. $l = 0$, 1 and 2), the restrictions for the chain (10.24) of all the allowed $SU(2l+1)$ representations are multiplicity-free.

In the case of the f^n ($l=3$) configurations the restriction is multiplicity-free for $n=0$, 1, and 2 only. It is at this point that Racah introduces G_2 into the chain (10.24) to form

$$SU(7) \supset SO(7) \supset G_2 \supset SO(3), \qquad (10.25)$$

which allows the treatment of f^3 and f^4 in a multiplicity-free way; the configurations f^n, $n=5$, 6 and 7, even with G_2 included, are not multiplicity-free and require an additional label.

Biographical Sketch

Racah, Gulio (1909–1965) was born in Florence, Italy. He received the Ph.D. from Florence University in 1930, studied in Rome with Fermi, emigrated to Palestine in 1939. He was appointed Professor of Theoretical Physics at Hebrew University of Jerusalem and later became Dean of the Faculty of Science, Rector and acting President. His major contributions were in the theory of angular momentum and atomic spectroscopy. His name is associated with the Racah coefficient (also known as the $6-j$ symbol) and the Racah–Wigner calculus.

11
Casimir operators for orthogonal groups

In Chapter 4 we introduced the quadratic Casimir operator as a quadratic polynomial formed out of the generators and commuting with all the generators. In this chapter we describe the generalization to polynomials of higher degree in the case of the orthogonal groups.

It should be noted that, in contrast to the quadratic Casimir operator, there is no *canonical* choice for the higher-degree Casimir operators. Below, we describe a particular solution to the problem.

Starting from the commutation relations (5.3)

$$[A_{ij}, A_{mn}] = i(\delta_{m[i}A_{j]n} - A_{m[i}\delta_{j]n}) \tag{11.1}$$

that define the orthogonal algebras we readily obtain

$$[A_{ij},(A^2)_{mn}] = [A_{ij}, A_{mp}A_{pn}] = [A_{ij}, A_{mp}]A_{pn} + A_{mp}[A_{ij}, A_{pn}]$$
$$= i(\delta_{m[i}A_{j]p} - A_{m[i}\delta_{j]p})A_{pn} + iA_{mp}(\delta_{p[i}A_{j]n} - A_{p[i}\delta_{j]n})$$
$$= i(\delta_{m[i}(A^2)_{j]n} - (A^2)_{m[i}\delta_{j]n}). \tag{11.2}$$

Here we think of A as an $n \times n$ "matrix" with "matrix elements" A_{ij} and define A^2 by the rules of matrix multiplication. It readily follows by induction on s that

$$[A_{ij}, (A^s)_{mn}] = i(\delta_{m[i}(A^s)_{j]n} - (A^s)_{m[i}\delta_{j]n}) \tag{11.3}$$

and therefore if we define

$$\mathrm{Tr}\, A^s = (A^s)_{kk} \tag{11.4}$$

then

$$[A_{ij}, \mathrm{Tr}\, A^s] = 0 \tag{11.5}$$

and we have the desired generalized Casimir operators C_s of degree s:

$$C_s = \text{Tr} A^s. \tag{11.6}$$

Now the question arises: how many of these C_s are independent? The answer is that the number of polynomially independent generalized Casimirs is equal to the rank of the group. For a semisimple group the **rank** can be defined as the maximal number of mutually commuting generators. These generators form an Abelian subalgebra known as the Cartan subalgebra. In the case of $so(2m)$ and $so(2m+1)$ it is clear that in the basis that we have been using the following m generators mutually commute: $A_{12}, A_{34}, \ldots, A_{2m-1,2m}$; moreover the remaining generators fail to commute with at least one of these m generators. Thus, for both $so(2m)$ and $so(2m+1)$ the rank is m.

To see which of the C_s are independent we first recall that the Cayley–Hamilton theorem for an ordinary $n \times n$ matrix M implies that $\text{Tr } M^{n+1}$ can be expressed in terms of traces of lower powers. Now our "matrix" A differs from an ordinary matrix in that its "matrix elements" do not commute. Since the commutator of two generators is proportional to a single generator, this lack of commutativity results in there possibly being corrections—terms of a given degree in the ordinary Cayley–Hamilton theorem might be accompanied by terms of *lower* degree. Consequently, it will still be true that for $so(n)$ $\text{Tr} A^{n+1}$ can be expressed in terms of traces of lower powers.

Secondly, recall that the generators are antisymmetric. For an ordinary antisymmetric matrix M we obviously have $\text{Tr } M^s = 0$ for odd s. For our "matrix" A this means that $\text{Tr } A^s$ for odd s can be expressed in terms of traces of lower powers.

Putting these results together we conclude that for both $so(2m)$ and $so(2m+1)$ the independent Casimirs are

$$C_{2k} = \text{Tr} A^{2k}, \quad k=1,2,\ldots,m. \tag{11.7}$$

This conclusion requires one modification. The existence of the invariant totally antisymmetric ε tensor means that for $so(2m)$ we can form another invariant out of the generators, namely

$$P = \varepsilon_{\underbrace{abcd\ldots uw}_{2m}} \overbrace{A_{ab} A_{cd} \ldots A_{uw}}^{m}, \tag{11.8}$$

which gives a Casimir of degree m. P is a Casimir because all the indices are summed over. We recall that we have defined a tensor of rank s as a collection of components $T_{ab...w}$ with s free indices such that the commutator $[A_{mn}, T_{ab...w}]$ has a specified form. In particular, when all the indices are summed over, leaving zero free indices, we have a tensor of rank 0 that commutes with the generators, i.e. is invariant under rotations— an excellent example of such an invariant is the length squared of an n-vector.

Now clearly P^2 is of degree $2m$. Thus, in the case of $so(2m)$ a polynomially independent set is obtained by omitting in (11.7) $k=m$ and including instead P. We further note that P is not an $O(m)$ invariant since under a reflection $P \to -P$. It follows that irreducible representations of $so(2m)$ for which the eigenvalue of P is non-zero have inequivalent twins as we already know—these are the semispinors.

We also recall that the tensor representations have no such twins, which implies that for them P must be zero. Indeed, we easily see this to be the case for the vector representation where $(A_{ab})_{pq} = -i\delta_{p[a}\delta_{b]q}$:

$$P = \varepsilon_{abcd...uw}(A_{ab})_{pq}(A_{cd})_{qr}...(A_{uw})_{mn}$$
$$= (-i)^m \varepsilon_{abcd...uw} \delta_{p[a}\delta_{b]q}\delta_{q[c}\delta_{d]r}...\delta_{m[u}\delta_{w]n}$$
$$= (-2i)^m \varepsilon_{pqr...mn} = 0. \qquad (11.9)$$

The fact that for $so(2m)$ we must include P in our basis for the Casimirs is glorified by the concept of an *integrity basis*. All Casimir operators can be obtained by forming polynomials out of the operators in the integrity basis, this being its definition. The point is that had we taken instead for our basis the Tr A^s, $s=2, 4, \ldots, 2m$, we *couldn't* produce P by forming polynomials out of the above—to obtain P in that way requires taking square roots with their $\pm$ options. It is to guard against this type of ambiguity that one needs an integrity basis.

For the record we note that the existence of P is related to the curious fact that the determinant of an antisymmetric $2m \times 2m$ matrix is a perfect square, the object that it is a square of being called a Pfaffian.

The Casimir invariant P also plays an important role in conjugation. We recall that if we have an irreducible d-dimensional representation with the generators given by some $d \times d$ matrices X_{ab} then the $d \times d$ matrices Y_{ab}, where

$$Y_{ab} = -X_{ab}^T \qquad (11.10)$$

describe another irreducible representation, the so-called conjugate representation, which may or may not be equivalent to the original one.

In the first representation the Casimir invariant P is given by

$$P = \varepsilon_{\underbrace{abcd...uw}_{2m}} \overbrace{X_{ab} X_{cd} ... X_{uw}}^{m}, \qquad (11.11)$$

which, by Schur's lemma can be set equal to $p 1_d$, where 1_d is the $d \times d$ unit matrix and p is some number. In the conjugate representation this Casimir invariant is given by

$$P^C = \varepsilon_{\underbrace{abcd...uw}_{2m}} \overbrace{Y_{ab} Y_{cd} ... Y_{uw}}^{m}$$

$$= (-1)^m \varepsilon_{abcd...uw} X_{ab}{}^T X_{cd}{}^T ... X_{uw}{}^T$$

$$= (-1)^m \varepsilon_{abcd...uw} (X_{uw} ... X_{cd} X_{ab})^T$$

$$= (-1)^m \varepsilon_{uw...cdab} (X_{uw} ... X_{cd} X_{ab})^T$$

$$= (-1)^m (p 1_d)^T = (-1)^m P \qquad (11.12)$$

so that for m odd the two representations are ineqivalent for $p \neq 0$, which is the case for the semispinors. These results are in agreement with our remark in Chapter 8 that complex representations are only possible if independent Casimir operators of odd degree exist.

Again, the existence of P is crucial since all the other Casimir operators of the form C_s are invariant under conjugation:

$$C_s{}^C = \overbrace{Y_{ab} Y_{bc} ... Y_{uw} Y_{wa}}^{s} = (-1)^s X_{ab}{}^T X_{bc}{}^T ... X_{uw}{}^T X_{wa}{}^T$$

$$= (-1)^s (X_{wa} X_{uw} ... X_{bc} X_{ab})^T = (-1)^{2s} (X_{aw} X_{wu} ... X_{cb} X_{ba})^T$$

$$= C_s{}^T = C_s. \qquad (11.13)$$

Biographical Sketch

Pfaff, Johann Friedrich (1765–1825) was born in Stuttgart, Germany. He studied in Göttingen, Berlin and Vienna. He became professor of mathematics at the University of Helmstedt, where Gauss attended his lectures and they became friends. When that university closed in 1810 he went to Halle where he died.

12
Classical groups

The **classical** groups consist of the orthogonal groups, on which we have spent so much time already, and two other kinds—the unitary and the symplectic groups.

This classification comes about by thinking about groups whose elements are *linear transformations* on some entities, usually called vector components, which keep invariant a certain *quadratic form*. Thus, suppose that we have n entities x_k, $k=1,2,...,n$, and we perform a linear transformation on them

$$x'_l = T_{lk} x_k, \qquad (12.1)$$

such that

$$x'_l x'_l = x_k x_k, \qquad (12.2)$$

which is ensured if T_{lk} satisfies

$$T_{lk} T_{lm} = \delta_{km}. \qquad (12.3)$$

In obvious matrix notation we may rewrite (11.1)–(11.3) as

$$x' = Tx, \quad x'^{\mathrm{T}} x' = x^{\mathrm{T}} x, \quad T^{\mathrm{T}} T = 1_n. \qquad (12.4)$$

Now, in general arbitrary $n \times n$ matrices T form a group under matrix multiplication provided det $T \neq 0$, the so-called **general linear** group. This group is denoted by $GL(n,\mathbb{R})$ or $GL(n,\mathbb{C})$ depending on whether the matrix elements are in $\mathbb{R}$ or in $\mathbb{C}$. We recognize the subgroup of $GL(n)$ obeying (12.3), i.e. satisfying

$$T^{\mathrm{T}} = T^{-1} \qquad (12.5)$$

as the **orthogonal** group $O(n,\mathbb{R})$ or $O(n,\mathbb{C})$.

However, we could be more general and make the quadratic form that the transformations leave invariant a bit more fancy by introducing what is generally called a *metric* M and requiring instead invariance of $x^T M x$, so that the matrices T have to satisfy instead

$$T^T M T = M, \qquad (12.6)$$

which, of course, reduces to the ordinary orthogonal case if M is the unit matrix.

Now, the kind of structure that results depends on this metric M. We consider first the situation when the x, and therefore T and M, are all in $\mathbb{R}$.

Suppose that M is *symmetric*—then it can be diagonalized and its eigenvalues are real. If we demand that the metric be non-singular then it can have no zero eigenvalues and so can be taken without loss of generality in the form of a diagonal matrix D with k entries -1, $n-k$ entries $+1$, $k \leq n-k$, $k = 0, 1, 2, ...$

$$D = \mathrm{diag}\Big(\underbrace{+1, +1, ..., +1}_{n-k}, \underbrace{-1, -1, ..., -1}_{k} \Big). \qquad (12.7)$$

This form is referred to as having *signature* $(\underbrace{+1, +1, ..., +1}_{n-k}, \underbrace{-1, -1, ..., -1}_{k})$.

What we have here are the **generalized orthogonal** groups denoted by

$$O(n-k, k, \mathbb{R}). \qquad (12.8)$$

These groups are in many ways like the ordinary orthogonal groups except for being non-compact for $k \neq 0$. When $k=0$ we are back, of course, to the ordinary orthogonal groups—they can be characterized by stating that they preserve a quadratic form with a symmetric positive definite metric.

Many of the groups of importance in Physics are generalized orthogonal groups. $O(4,1,\mathbb{R})$ and $O(3,2,\mathbb{R})$, the so-called De Sitter and anti-de Sitter group, are important in general relativity; $O(4,2,\mathbb{R})$, the so-called Liouville or *conformal* group is the group that leaves invariant Maxwell's equations. Perhaps the most important is $O(3,1,\mathbb{R})$, the so-called Lorentz group of special relativity, the group that leaves invariant the interval between two events in 3+1 space-time. The six generators of this group correspond to the three ordinary rotations in the three space dimensions and the three "rotations" in space-time planes, the so-called **boosts**. A rotation followed by a rotation is again a rotation because $O(3,\mathbb{R})$ is a subgroup of $O(3,1,\mathbb{R})$. But a boost followed by another boost is in general not a boost (except if the two boosts are collinear) but a boost and a rotation—this is the reason for the phenomenon known as *Thomas precession*. The six parameters of this group can be identified with the three Euler angles needed to specify an arbitrary rotation in the three space dimensions and the three components of the velocity between the two frames in constant relative motion.

Next, suppose that M is *antisymmetric*. Since

$$\det M = \det M^\mathrm{T} = (-1)^n \det M \tag{12.9}$$

it follows that M is *singular* if n is odd so we assume that $n=2m$.

By definition, a **symplectic** transformation S is a $2m \times 2m$ matrix satisfying

$$S^\mathrm{T} J S = J, \quad \text{where} \quad J^\mathrm{T} = -J. \tag{12.10}$$

A canonical form for J (which cannot be diagonalized in $\mathbb{R}$) is

$$J = \operatorname{diag}(\underbrace{j,j,...j}_{m}), \quad j = \begin{pmatrix} 0 & 1 \\ -1 & 0 \end{pmatrix}. \tag{12.11}$$

These matrices S form the *symplectic* group $Sp(2m,\mathbb{R})$, which is non-compact.

Next suppose that $x \in \mathbb{C}$, therefore the matrices and the metric are all in $\mathbb{C}$.

For M symmetric we can take it without loss of generality to be **1**. This is because were we to take $M=D$ so that the matrices T would satisfy $T^\mathrm{T} D T = D$ we could then form the matrices T' that satisfy $T'^\mathrm{T} T' = 1$, where $T' = PTP$, with $P = \operatorname{diag}(\underbrace{+1,+1,...+1}_{n-k},\underbrace{i,i,...i}_{k})$. So we have $O(n,\mathbb{C})$.

This group is non-compact.

For $M=J$ (*antisymmetric*) we have the symplectic group $Sp(2m,\mathbb{C})$.

The quadratic form $x^\mathrm{T} M x$ considered so far is called **bilinear**. When the vectors x and their transformations T are in $\mathbb{C}$ we could also have a so-called **sesquilinear** quadratic form given by

$$x^\dagger M x, \tag{12.12}$$

where the dagger denotes hermitian conjugate (i.e. complex conjugate transpose). Now the matrices U that preserve this quadratic form satisfy

$$U^\dagger M U = M \qquad (12.13)$$

and for $M=D$ the matrices U form the **generalized unitary** groups denoted by

$$U(n-k,k,\mathbb{C}). \qquad (12.14)$$

For $k \neq 0$ these groups are non-compact, for $k=0$ we have the ordinary unitary group, which is compact. We note that obviously

$$U(n-k,k,\mathbb{R}) = O(n-k,k,\mathbb{R}) \qquad (12.15)$$

since the sesquilinear and bilinear quadratic forms are indistinguishable in $\mathbb{R}$. It is customary to denote $U(n-k,k,\mathbb{C})$ as simply $U(n-k,k)$ and $O(n-k,k,\mathbb{R})$ as simply $O(n-k,k)$.

Matrices that are simultaneously unitary and symplectic form the **unitary symplectic** group denoted by $USp(2m-2k,2k)$:

$$USp(2m-2k,2k) = U(2m-2k,2k) \cap Sp(2m,\mathbb{C}) \qquad (12.16)$$

and these are compact for $k=0$, non-compact for $k \neq 0$.

As mentioned before the classical groups are defined in terms of matrix groups preserving quadratic forms. We mention a few more matrix groups. The general linear groups over $\mathbb{R}$ and $\mathbb{C}$ have obvious subgroups, called **special linear** groups and denoted by $SL(n,\mathbb{R})$ and $SL(n,\mathbb{C})$, defined by the requirement that they be unimodular. $GL(n,\mathbb{C})$ has in addition the subgroups $SL_1(n,\mathbb{C})$ with real determinant and $SL_2(n,\mathbb{C})$ with determinant of modulus unity.

Two more groups need be mentioned. The matrices in $SO(2m,\mathbb{C})$, which leave invariant an antisymmetric sesquilinear form, are a group denoted by $SO^*(2m)$. The matrices in $SL(2m,\mathbb{C})$, which commute with the product of an antisymmetric matrix and the complex conjugation operator, form a group denoted by $SU^*(2m)$.

In addition to matrices with real or complex matrix elements we could also discuss matrices with quaternionic matrix elements, taking care to account for the non-commutativity of quaternions. This allows one to define the classical groups in a unified way as follows. Let $U(n-k,k,\mathbb{F})$ be the *unitary* group whose elements are $n \times n$ matrices with matrix elements in a field $\mathbb{F}$, which preserve the quadratic form $q^\dagger D q$, where

$$q = \begin{pmatrix} q_1 \\ q_2 \\ \cdot \\ \cdot \\ q_n \end{pmatrix} \qquad (12.17)$$

with $q_j \in \mathbb{F}$. The field $\mathbb{F}$ can be $\mathbb{R}$, $\mathbb{C}$ or $\mathbb{H}$ and the dagger denotes, respectively, transpose, transpose complex conjugate and transpose quaternionic conjugate.

In view of the isomorphism between quaternions and $su(2)$ (see Chapter 9) an $n \times n$ matrix with quaternionic elements can be viewed as an $2n \times 2n$ matrix with complex elements, thus avoiding non-commutativity issues. In particular, one can show that this leads to the isomorphism

$$U(n-k,k,\mathbb{H}) \cong USp(2n-2k,2k). \qquad (12.18)$$

Thus, the unitary groups over $\mathbb{F}$ are, respectively, the orthogonal ($\mathbb{F}=\mathbb{R}$), unitary ($\mathbb{F}=\mathbb{C}$) and unitary symplectic ($\mathbb{F}=\mathbb{H}$) groups.

We use this unified approach to find the dimensions and describe the topology of $U(n,\mathbb{F})$. Consider then $n \times n$ matrices U with matrix elements in $\mathbb{F}$, which obey

$$U^\dagger U = \mathbf{1}. \qquad (12.19)$$

In detail, let U be given as

$$U = \begin{pmatrix} a_1 & b_1 & c_1 & \ldots & w_1 \\ a_2 & b_2 & c_2 & \ldots & w_2 \\ a_3 & b_3 & c_3 & \ldots & w_3 \\ \cdot & \cdot & \cdot & \ldots & \cdot \\ a_n & b_n & c_n & \ldots & w_n \end{pmatrix} \qquad (12.20)$$

so that (12.19) stands for the following equations:

$$a_k^* a_k = 1,$$
$$b_k^* a_k = 0,\ b_k^* b_k = 1,$$
$$c_k^* a_k = 0,\ c_k^* b_k = 0,\ c_k^* c_k = 1,$$
$$\text{etc.} \qquad (12.21)$$

In these equations we may think of a as a vector with n components $a_k \in \mathbb{F}$, i.e. $a \in \mathbb{F}^n$. Thus, respectively, in the three cases we have

$a \in \mathbb{R}^n$, $a \in \mathbb{C}^n = \mathbb{R}^{2n}$ and $a \in \mathbb{H}^n = \mathbb{R}^{4n}$ and the same is true of the vectors b,c, etc.

Now the first equation in (12.21), $a_k^* a_k = 1$, is a *real* equation because of the way the star operation is defined. It therefore constitutes *one* constraint on the n entries in the first column of U and so that column can be parameterized by $fn-1$ real parameters with $f=1,2$ and 4, respectively, in the case of $O(n)$, $U(n)$ and $USp(2n)$; in other words this constraint tells us that the first column of U defines the sphere S^{fn-1}.

The vectors $a,b,c,...$ have the indicated components with respect to some set of n orthogonal axes. We can rotate those axes so that with respect to the rotated axes all the components of a are zero except for a_1. Then the requirement that a be orthogonal to $b^*, c^*,...$ is satisfied if $0 = b_1 = c_1 = ...$ The remaining requirement in the second line of (12.21), $b_k^* b_k = 1$, defines now $S^{f(n-1)-1}$ as there are only $n-1$ non-vanishing components in b.

We can now contemplate rotating the axes in the $(n-1)$-dimensional subspace that does not include the axis labeled 1, such that with respect to the rotated axes all components of b are zero except for b_2. Now the requirement that b be orthogonal to $c^*,...$ is satisfied if $0 = c_2 = ...$ The remaining requirement in the third line of (12.21), $c_k^* c_k = 1$, defines now $S^{f(n-2)-1}$ as there are only $n-2$ non-vanishing components in c.

Continuing in this fashion we find that the number of parameters is

$$\sum_{k=1}^{n}(fk-1) = (f\tfrac{n+1}{2}-1)n \xrightarrow[f=1]{} n(n-1)/2 \quad \text{for} \quad O(n)$$

$$\xrightarrow[f=2]{} n^2 \quad \text{for} \quad U(n) \quad (12.22)$$

$$\xrightarrow[f=4]{} n(2n+1) \quad \text{for} \quad USp(2n)$$

and that the topology is

$$\bigcup_{k=1}^{n} S^{fk-1} \quad (12.23)$$

and we remark that all $S^n, n \geq 2$, are simply connected, while S^1 is infinitely connected.

Thus, for the case of the orthogonal groups we see that $SO(2)$ is S^1 and so infinitely connected, while for $SO(3)$ we get the union of S^1 and S^2. But we know a different parameterization for $SO(3)$ that makes its topology

Classical groups 99

that of a three-dimensional ball with antipodal points identified, showing that $SO(3)$ is doubly connected. If then for $n\geq 3$ we stop the induction from $SO(n)$ down at $SO(3)$ we have the topology of the union of $\mathbb{S}^{n-1}$, $\mathbb{S}^{n-2}$, ..., $\mathbb{S}^4$ and the doubly connected three-dimensional ball, hence the $SO(n)$ for $n\geq 3$ are doubly connected.

For $U(n)$ we get from (12.23) the union of $\mathbb{S}^{2n-1}$, $\mathbb{S}^{2n-3}$, ..., $\mathbb{S}^3$, $\mathbb{S}^1$ and so, because of $\mathbb{S}^1$, the $U(n)$ groups are not simply connected. As a result of the rotations described above our matrix U (12.20) is in diagonal form: $\text{diag}(e^{i\varphi_1}, e^{i\varphi_2},...,e^{i\varphi_n})$ where we are free to choose the phases as we like except that $\exp i(\varphi_1+\varphi_2+...+\varphi_n)=\det U$. Consequently, for $SU(n)$ we can choose all the phases to be zero and that eliminates the $\mathbb{S}^1$ from the union of the spheres so that $SU(n)$ is simply connected.

For $USp(2n)$ we get from (12.23) the union of $\mathbb{S}^{4n-1}$, $\mathbb{S}^{4n-5}$,..., $\mathbb{S}^7$, $\mathbb{S}^3$. All these spheres are simply connected and therefore $USp(2n)$ is simply connected.

In conclusion, we list the dimensions of all the matrix groups discussed in this chapter:

Group	Dimension
$GL(n,\mathbb{R})$	n^2
$GL(n,\mathbb{C})$	$2n^2$
$SL(n,\mathbb{R})$	n^2-1
$SL(n,\mathbb{C})$	$2(n^2-1)$
$SL_1(n,\mathbb{C})$	$2n^2-1$
$SL_2(n,\mathbb{C})$	$2n^2-1$
$O(n-k,k)$	$n(n-1)/2$
$O(n,\mathbb{C})$	$n(n-1)$
$SO^*(2m)$	$m(2m-1)$
$Sp(2m,\mathbb{R})$	$m(2m+1)$
$Sp(2m,\mathbb{C})$	$2m(2m+1)$
$USp(2m-2k,2k)$	$m(2m+1)$
$U(n-k,k)$	n^2
$SU(n-k,k)$	n^2-1
$SU^*(2m)$	$(2m)^2-1$.

This list can be used as a guide for possible isomorphisms or homomorphisms among matrix groups, such as e.g. among $O(3,1)$, $O(3,\mathbb{C})$ and

$SL(2,\mathbb{C})$, which can all be used to describe the Lorentz group. Another example is $O(2n+1)$ and $USp(2n)$ that are isomorphic for n equal to one and two.

Biographical Sketches

Lorentz, Hendrik Anton (1853–1928) was a Dutch physicist born in Arnhem. At age 25 he was offered the first Chair of Theoretical Physics at Leiden, which had been created for van der Waals. He made major contributions with his electron theory. In 1902 he shared with Zeeman the Nobel prize for Physics. In 1904 he derived a mathematical transformation, the Lorentz–Fitzgerald contraction, which explained the apparent absence of relative motion between the Earth and the ether.

de Sitter, Willem (1872–1934) was a Dutch astronomer born in Sneek, Friesland. He was appointed Director and Professor of Astronomy at the University of Leiden in 1908. After Einstein had solved his equations of general relativity to produce a description of a static universe with curved space, the curvature being constant in time, de Sitter demonstrated that an expanding universe of constantly decreasing curvature emerged as another solution.

Liouville, Joseph (1809–82) was a French mathematician born in St. Omer. He taught at École Polytechnique (1831–51) and then at the Collége de France and the University of Paris. In 1836 he founded and edited for nearly 40 years the *Journal de Mathématiques*. He worked in analysis, theory of differential equations, mathematical physics and celestial mechanics. Among his contributions to number theory is the proof that transcendental numbers are infinitely many.

Maxwell, James Clerk (1831–79) was a Scottish physicist born in Edinburgh. From 1860 to 1865 he was Professor of Natural Philosophy and Astronomy at King's College, London. In 1871 he was appointed the first Cavendish Professor of Experimental Physics at Cambridge. He is regarded as one of the founders of the kinetic theory of gases. Out of this work came the statistical interpretation of thermodynamics and the idea of the Maxwell demon, which would appear to violate the second law of thermodynamics. In "A Dynamical Theory of the Electromagnetic Field"

(1864) Maxwell put forward four differential equations now famous and bearing his name.

Thomas, Llewellyn (1903–92) was born in London and studied at Cambridge University where he received the Ph.D. degree in 1928. He was Professor of Physics at Ohio State University (1929–43), staff member at Watson Scientific Computing Laboratory at Columbia University (1946–68), and Visiting Professor at North Carolina State University (1968–76). He is best known for Thomas precession and the Thomas–Fermi statistical model of the atom.

13
Unitary groups

By definition, the $U(n)$ group consists of all unitary complex $n \times n$ matrices with multiplication defined as matrix multiplication. A matrix U is unitary if its inverse equals its hermitian conjugate (=complex conjugate transpose). As usual, the $SU(n)$ group is defined as the unimodular subgroup of $U(n)$. It follows from the unitarity condition

$$U^\dagger U = 1 \qquad (13.1)$$

that all matrix elements of U have magnitude bounded by unity so the volume in parameter space is finite and the $U(n)$ group is compact. In the preceding chapter we have shown that the dimension of $U(n)$ is n^2 and so an element of $U(n)$ in the neighborhood of unity can be parameterized as

$$\exp(i\varphi_\alpha F_\alpha), \quad \alpha=1,2,\ldots,n^2, \qquad (13.2)$$

where the φ_α are the parameters and the F_α are the generators, which are required to be hermitian in a unitary representation.

The F_α generate the $u(n)$ Lie algebra. Consider for a moment $n \times n$ regular *real* matrices—they form the group $GL(n,\mathbb{R})$, the group of general linear transformations on a n-plet of real numbers. It is obvious that we have a basis in terms of the n^2 quantities $E_a{}^b$, a, $b=1, 2,\ldots,n$, which are $n \times n$ matrices with zero entries everywhere except for unity at the intersection of the ath row and bth column:

$$(E_a{}^b)_s{}^t = \delta_{as}\delta^{bt}. \qquad (13.3)$$

Any real $n \times n$ matrix can be written as $r_b{}^a E_a{}^b$, $r_b{}^a$ being n^2 real parameters. The algebra $\mathrm{gl}(n,\mathbb{R})$ is thus defined by

$$\begin{aligned}
([E_a{}^b, E_c{}^d])_s{}^t &= (E_a{}^b)_s{}^k (E_c{}^d)_k{}^t - (E_c{}^d)_s{}^k (E_a{}^b)_k{}^t \\
&= \delta_{as}\delta^{bk}\delta_{ck}\delta^{dt} - \delta_{cs}\delta^{dk}\delta_{ak}\delta^{bt} \\
&= \delta_c{}^b (E_a{}^d)_s{}^t - (E_c{}^b)_s{}^t \delta_a{}^d,
\end{aligned} \qquad (13.4)$$

from which we abstract the commutation relations

$$[E_a{}^b, E_c{}^d] = \delta_c{}^b E_a{}^d - E_c{}^b \delta_a{}^d. \tag{13.5}$$

Now, for the $u(n)$ Lie algebra we want n^2 *hermitian* generators, while the $E_a{}^b$ defined above are not hermitian for $a \neq b$. But clearly we can form for $a \neq b$

$$K_{ab} = E_a{}^b + E_b{}^a \quad \text{and} \quad L_{ab} = \mathrm{i}(E_a{}^b - E_b{}^a), \tag{13.6}$$

where K_{ab} and L_{ab} are hermitian. The point of this exercise is to indicate that we can deal just as well with the non-hermitian $E_a{}^b$, i.e. the Lie algebras of $U(n)$ and $GL(n,\mathbb{R})$ are isomorphic:

$$u(n) \cong gl(n,\mathbb{R}). \tag{13.7}$$

Next, we observe that it follows from the unitarity condition (13.1) that $|\det U|^2 = 1$ and therefore

$$\det U = \exp\mathrm{i}\varphi, \quad \varphi \text{ real}, \tag{13.8}$$

so the unimodularity constraint reduces the number of parameters by just one. Thus, elements of $SU(n)$ can be parameterized as in (13.2) but by $n^2 - 1$ parameters and with the F_α required to be traceless. Since $E_k{}^k = 1$ it follows that $E_k{}^k$ generates a $U(1)$ subgroup whose elements commute with everything and the $n^2 - 1$ traceless generators of $SU(n)$ can be taken as

$$E'_a{}^b = E_a{}^b, \quad a \neq b,$$
$$E'_a{}^a = E_a{}^a - \tfrac{1}{n}\mathbf{1}, \quad \text{(no sum on } a!\text{)} \tag{13.9}$$

and the primed generators obey the same commutation relations as the unprimed ones.

By the way, it follows that $U(n)$ is *not* semisimple since it contains the "$U(1)$ factor". $SU(n)$ is semisimple.

Next we ask: what is the center of $SU(n)$? Well, we need an $n \times n$ matrix that commutes with everything—$\lambda \mathbf{1}_n$, and if it is to be unimodular then we must have $\lambda^n = 1$. The solution of this equation in $\mathbb{C}$ yields the nth roots of unity, so the center is $\mathbb{Z}_n$. Thus, in $SU(3)$ the center is $\mathbb{Z}_3$ and this is why there are *three* kinds of representations: the quark **3**, the antiquark **3*** and the meson **8** being examples of each. Since the nth roots of unity are complex for $n>2$ it follows that *complex* representations occur. For $n=2$ this argument does not apply since the two square roots of unity are ± 1 and *real*. Indeed we know, in view of the isomorphism of $SU(2)$ and $Spin(3)$, that the irreducible representations of $SU(2)$ are self-conjugate: real for integer spin and symplectic for half-odd-integer spin. This property of $SU(2)$ was called by Wigner **ambivalence**.

That complex conjugation is an automorphism of $U(n)$ and $SU(n)$ follows from our commutation relations (13.5). For $n=2$ complex conjugation can be accomplished by a similarity transformation. For $n>2$ that is, in general, not possible, as can be seen by considering the center. In some representations the element that generates the center is represented by **1** (for n odd) or by **1** or **−1** (for n even) and these representations are self-conjugate. In all other representations that element is represented by **1** times a complex nth root of unity, therefore its trace is complex and cannot be transformed into its complex conjugate by a similarity transformation.

Just as for the orthogonal groups we can develop the concept of tensor operators and use that to find Casimir operators. With the realization of the generators in the form

$$E_a{}^b = z_a \partial^b, \quad \partial^b \equiv \partial/\partial z_b \tag{13.10}$$

we find

$$[E_a{}^b, z_c] = \delta_c{}^b z_a, \quad [E_a{}^b, \partial^c] = -\delta_a{}^c \partial^b, \tag{13.11}$$

which suggests that we define two types of rank one tensors that transform like z_a or ∂^a. So we define a **contravariant tensor of rank one** to be a collection of n entities T_a, $a=1,2,...,n$, which obey

$$[E_a{}^b, T_c] = \delta_c{}^b T_a \tag{13.12}$$

and a **covariant tensor of rank one** to be a collection of n entities T^a, $a=1,2,..., n$, which obey

$$[E_a{}^b, T^c] = -\delta_a{}^c T^b. \tag{13.13}$$

Now, a **tensor of rank two** comes in three varieties: *covariant* T^{ab}, which consists of n^2 entities obeying

$$[E_a{}^b, T^{cd}] = -\delta_a{}^c T^{bd} - \delta_a{}^d T^{cb}, \tag{13.14}$$

contravariant T_{ab}, which consists of n^2 entities obeying

$$[E_a{}^b, T_{cd}] = \delta_c{}^b T_{ad} + \delta_d{}^b T_{ca}, \tag{13.15}$$

and *mixed* $T_a{}^b$, which consists of n^2 entities obeying

$$[E_a{}^b, T_c{}^d] = \delta_c{}^b T_a{}^d - \delta_a{}^d T_c{}^b. \tag{13.16}$$

These tensors are reducible. We have

$$T^{ab} = (T^{(ab)} + T^{[ab]})/2, \quad n^2 = n(n+1)/2 + n(n-1)/2, \tag{13.17}$$

and similarly for T_{ab}. The mixed $T_a{}^b$ reduces differently:

$$T_a{}^b = \tfrac{1}{n}\delta_a{}^b T + \left(T_a{}^b - \tfrac{1}{n}\delta_a{}^b T\right), \quad n^2 = 1 + (n^2 - 1), \tag{13.18}$$

where $T \equiv T_a{}^a$ is a tensor of rank zero, an **invariant**. The reason for reducibility of T^{ab} or T_{ab} is due to the fact that the symmetrized and antisymmetrized parts transform separately, as will be demonstrated in the next chapter. However, we cannot symmetrize or antisymmetrize a superscript and a subscript and the reducibility of $T_a{}^b$ is due to the existence of the *invariant* mixed tensor $\delta_a{}^b$:

$$[E_a{}^b, \delta_c{}^d] = \delta_c{}^b \delta_a{}^d - \delta_a{}^d \delta_c{}^b = 0. \tag{13.19}$$

Observe that $E_a{}^b$, the generators of $U(n)$, are precisely a *mixed* second rank tensor, which break up into the invariant $E_a{}^a$ and the $n^2 - 1$ generators $E'_a{}^b$ of $SU(n)$. Clearly we thus have that $E_a{}^a \equiv \operatorname{Tr} E$ is a **linear Casimir** operator of $U(n)$ and we can form a **quadratic Casimir**

$$C_2 = E_a{}^b E_b{}^a \equiv \operatorname{Tr} E^2, \tag{13.20}$$

and a **cubic Casimir**

$$C_3 = E_a{}^b E_b{}^c E_c{}^a \equiv \operatorname{Tr} E^3, \tag{13.21}$$

etc., where we denote by E the $n \times n$ "matrix" whose ab "matrix element" is $E_a{}^b$. The proof that the trace of powers of E is a Casimir proceeds in complete analogy to the case of the orthogonal groups as discussed in Chapter 11.

Again as discussed for the orthogonal groups it follows from the Cayley–Hamilton theorem that $\operatorname{Tr} E^{n+1}$ is not polynomially independent of the traces of lower powers and therefore an integrity basis for the Casimir operators can be chosen in the form

$$s=1,2,\ldots,n \text{ for } U(n)$$
$$C_s = \mathrm{Tr} E^s \qquad (13.22)$$
$$s=2,3,\ldots,n \text{ for } SU(n),$$

the number of independent Casimirs being equal to the rank.

Note that $SU(n)$, $n>2$, contains in the integrity basis for Casimir operators a *cubic* Casimir, and so in general gives rise to anomalies. The *anomaly* is a property of a representation that can be shown to be proportional to the cubic Casimir of the representation. Consistency of the quantum field theories for elementary particles requires that the representation used to describe the right-handed spin $1/2$ particles be anomaly-free. Now consider the $SU(3) \times SU(2) \times U(1)$ theory, the so-called Standard Model, or equivalently its unified into $SU(5)$ version. Here, the elementary particles are grouped in (three) families and in each family we have 15 right-handed spin $1/2$ particles to consider: in the first family we have the six quarks (in two flavors and three colors), six antiquarks, the electron, the positron and the electron antineutrino. There is no 15-dimensional irreducible representation in $SU(5)$ and so Georgi and Glashow considered in their theory the reducible **5**⊕**10*** representation. Remarkably, **5** and **10*** have equal and opposite cubic Casimirs. Having a reducible representation for the fundamental particles of the theory is certainly not very pleasing esthetically but the anomaly cancellation was quite remarkable and reassuring, although mysterious.

The mystery is explained (and the reducibility eliminated) when the unifying group is enlarged from $SU(5)$ to $Spin(10)$. In $Spin(10)$ each family is put into the spinor **16**, which decomposes with respect to its $SU(5)$ subgroup as

$$\mathbf{16} = \mathbf{5} \oplus \mathbf{10^*} \oplus \mathbf{1}. \qquad (13.23)$$

(The **1** contains the right-handed neutrino, so $Spin(10)$, in contrast to $SU(5)$, can accommodate a massive neutrino, which necessarily contains both handedness states.) Since the $SU(5)$ singlet has zero value for the cubic Casimir (a singlet has zero value for all Casimirs) the cancellation of the cubic Casimir's value for the **5** against the **10*** is explained since the $Spin(10)$ has no cubic Casimirs in its integrity basis.

14
The symmetric group S_r and Young tableaux

We go back to our discussion of tensors in $SU(n)$. We have learned that locating indices up or down makes a difference and that contracting an upper and a lower index reduces a tensor, i.e. that $\delta_b{}^a$ is an invariant tensor. In addition, we have in $SU(n)$ the invariant tensors $\overbrace{\varepsilon^{ab...pr}}^{n}$ and $\underbrace{\varepsilon_{ab...pr}}_{n}$, which can be used to raise or lower indices. Thus, e.g. if $T_{ab...p}$ is a contravariant tensor of rank $n-1$ then W^r defined by

$$W^{\mathrm{r}} \equiv \varepsilon^{ab...pr} T_{ab...p} \qquad (14.1)$$

is a covariant tensor of rank one.

As discussed in Chapter 13 for a rank one tensor we have the n entities T_a, $a=1,2,...,n$. For a rank two tensor we have the n^2 entities T_{ab} or the irreducible $n(n+1)/2$ entities $T_{(ab)} \equiv T_{ab}+T_{ba}$ and the $n(n-1)/2$ entities $T_{[ab]} \equiv T_{ab}-T_{ba}$. It is worth remarking that we *cannot* use δ_{ab} to further reduce $T_{(ab)}$ because the *invariant* tensor is the mixed $\delta_a{}^b$, not δ_{ab} (this is to be contrasted with the situation for the $SO(n)$ groups).

Explicit calculation shows that the $T_{(ab)}$ and $T_{[ab]}$ transform independently and do not mix with each other, justifying our claim of reducibility, and the question is: why?

Consider the **symmetric group** S_2 (also called the **permutation** group) on two objects, namely the two indices of T_{ab}. The action of the two elements of S_2, the identity e and the exchange p, on T_{ab} is:

$$eT_{ab}=T_{ab}, \quad pT_{ab}=T_{ba} \qquad (14.2)$$

and the multiplication table has the obvious structure

$$e^2=e, \quad p^2=e, \quad ep=pe=p. \qquad (14.3)$$

108 Lie Groups and Lie Algebras: A Physicist's Perspective

We can write

$$T_{(ab)}=(e+p)T_{ab}, \quad T_{[ab]}=(e-p)T_{ab} \qquad (14.4)$$

and note that $(e\pm p)$ are *eigenstates* of p to the eigenvalue ± 1.

If we denote by $U_a{}^b$ an element of our $n\times n$ unitary matrices in $U(n)$ then the statement that T transforms as a second rank tensor can also be written as

$$T'_{ab}=U_a{}^k U_b{}^l T_{kl}, \qquad (14.5)$$

hence

$$pU_a{}^k U_b{}^l T_{kl}=pT'_{ab}=T'_{ba}=U_b{}^s U_a{}^t T_{st}=U_a{}^t U_b{}^s T_{st}=U_a{}^t U_b{}^s p T_{ts}, \qquad (14.6)$$

i.e. the actions of the symmetric group S_2 and the unitary group $U(n)$ *commute*.

This was perhaps to be expected as they involve different qualities of the tensor T. But now if we have an eigenstate of p to, say, the eigenvalue -1 then we still must have an eigenstate of p to the eigenvalue -1 *after* the action of $U(n)$—and so T must be *reducible* into pieces corresponding to eigenstates of p.

We next consider the tensor of rank three T_{abc} and the symmetric group on three objects S_3. We remark first that S_3 has $6=3!$ elements. In general, consider S_r—how do we rearrange r objects? Well, we can take the first object and put it first, or second, or third, etc. i.e. in r different places. Having placed it we can place the second object in $r-1$ different places, and so on. Thus, we arrive at

$$r(r-1)(r-2)...1=r! \qquad (14.7)$$

different rearrangements, the elements of S_r.

Explicitly, the six elements of S_3 can be identified as follows:

$$\begin{aligned}
(1)(2)(3)&=e & eT_{abc}&=T_{abc} \\
(12)(3)&=p_{12} & p_{12}T_{abc}&=T_{bac} \\
(13)(2)&=p_{13} & p_{13}T_{abc}&=T_{cba} \\
(1)(23)&=p_{23} & p_{23}T_{abc}&=T_{acb} \\
(123)&=p_{123} & p_{123}T_{abc}&=T_{cab} \\
(132)&=p_{132} & p_{132}T_{abc}&=T_{bca}.
\end{aligned} \qquad (14.8)$$

This **cycle** notation is useful. (1)(2)(3) means that we have just one-cycles, the symbol (1) denotes the rearrangement that takes the first object and puts it in the first place, the symbol (2) denotes the rearrangement that takes the second object and puts it in the second place, the symbol (3) denotes the rearrangement that takes the third object and puts it in the third place—i.e. *no rearrangement*, which is why (1)(2)(3)=e= identity.

Next, (12)(3) means that we have the two-cycle (12) that takes the first object and puts it in the second place, takes the second object and puts it in the first place, this followed by the one-cycle (3) that leaves the third object in the third place. Similarly for (13)(2) (exchange first and third object leaving second unchanged) and (1)(23). Note that the *square of a two-cycle* is equal to e. It is clear that the one-cycles could be omitted and from now on we write just (12) for (12)(3), etc.

Lastly, we have three-cycles: the three-cycle (123) means permute 1, 2 and 3 *cyclically*, i.e. take the first object and put it in second place, take the second object and put it in third place, take the third object and put it in first place. Similarly for (132). Note that the square of a three-cycle is not the identity, but the cube is.

Note that the entire group S_r can be generated by just the two-cycles (ij), the **transposition** or exchange of the ith and jth entries, in fact just the *adjacent* two-cycles $(i,i+1)$, which are $r-1$ in number.

In particular, for S_3 we can use the two adjacent two-cycles (12) and (23) to obtain the remaining four elements of S_3 by forming appropriate products:

$$e=(12)(12)=(23)(23)$$
$$(13)=(12)(23)(12)=(23)(12)(23)$$
$$(123)=(12)(23)$$
$$(132)=(23)(12). \qquad (14.9)$$

Going back to S_2 we observe that the objects S, called the **symmetrizer**, and A, called the **antisymmetrizer**, defined by

$$S=\tfrac{1}{2}(e+p), \quad A=\tfrac{1}{2}(e-p), \quad e=(1)(2), \quad p=(12) \qquad (14.10)$$

are idempotent

$$S^2=S, \quad A^2=A, \qquad (14.11)$$

complete
$$S+A=e, \qquad (14.12)$$
and orthogonal
$$SA=AS=0, \qquad (14.13)$$
and are eigenvectors of the permutation $p=(12)$ to the eigenvalues ± 1
$$pS=S, \quad pA=-A. \qquad (14.14)$$
It is easy to verify that for S_3 S and A defined by
$$\begin{aligned} S&=\frac{1}{3!}\{e+(123)+(132)+(12)+(13)+(23)\} \\ A&=\frac{1}{3!}\{e+(123)+(132)-(12)-(13)-(23)\} \end{aligned} \qquad (14.15)$$
are idempotent and orthogonal, but *not* complete. They also satisfy (14.14) with $p=(12)$ or (13) or (23). This can be seen as follows: because we have a group we must have $pS=S$ because S is the sum of all elements and p on every element in turn produces a permutation of all elements. Virtually the same argument works for pA if we note that A has all even permutations occurring with a plus sign, all odd permutations with a minus sign, where the **even** permutations are e, (123) and (132), being given by a product of an even number of transpositions and the **odd** permutations are (12), (13) and (23), being given by an odd number of transpositions. Hence, the action of p, a single transposition, on A will result in the appearance of a minus sign.

In fact, we can immediately conclude that for S_r we can form
$$S=\frac{1}{r!}\sum_R R, \qquad (14.16)$$
where R stands for any permutation and we sum over all $r!$ of them. This object is the symmetrizer and we have by the same argument as above that
$$pS=S, \quad p=\text{any two-cycle} \qquad (14.17)$$
and therefore
$$RS=S, \qquad (14.18)$$

since any permutation R can be written as a product of two-cycles. From these results it is easy to prove idempotency of S:

$$S^2 = \frac{1}{r!}\sum_R RS = \frac{1}{r!}\sum_R S = S\left(\frac{1}{r!}\sum_R 1\right) = S. \tag{14.19}$$

We may similarly deduce that the antisymmetrizer A defined by

$$A = \frac{1}{r!}\sum_R \delta_R R \tag{14.20}$$

satisfies

$$pA = -A, \quad p = \text{any two-cycle}. \tag{14.21}$$

Here, δ_R is the parity of the permutation R: it is $+1$ if R is even, -1 if R is odd. It follows that

$$RA = \delta_R A \tag{14.22}$$

and we easily prove idempotency of A, as well as orthogonality of A and S.

These generalizations from S_2 to S_r are quite straightforward—but what about (14.12), the completeness statement? Clearly S and A do not add up to the identity in S_r, $r > 2$. In other words, whereas in S_2 there exist only *two* kinds of symmetries: symmetric and antisymmetric, in S_r there must exist more symmetry types. So consider in S_3, for example, symmetrizing in index 1 and 2 followed by antisymmetrizing in index 1 and 3. We get

$$[e-(13)][e+(12)] = e+(12)-(13)-(123) \equiv \varphi. \tag{14.23}$$

To see what we have we act on φ with the two two-cycles that generate the whole group:

$$(12)\varphi = e+(12)-(132)-(23) \equiv \chi, \tag{14.24}$$
$$(23)\varphi = (23)+(132)-(123)-(13) = \varphi-\chi, \tag{14.25}$$

that is φ and χ form a basis, that is what we have here is a *two-dimensional* representation of the symmetric group S_3.

In fact, explicit calculation shows that Y and Y' defined by

$$Y = \frac{1}{3}[e-(13)][e+(12)] = \frac{1}{3}[e+(12)-(13)-(123)], \qquad (14.26)$$

$$Y' = \frac{1}{3}[e-(12)][e+(13)] = \frac{1}{3}[e-(12)+(13)-(132)], \qquad (14.27)$$

are idempotent, mutually orthogonal and orthogonal to S and A, and we have the completeness statement in the form

$$S+A+Y+Y'=e. \qquad (14.28)$$

We remark that the choice of Y and Y' is not unique, we could have instead taken $\frac{1}{3}[e-(13)][e+(23)]$ and $\frac{1}{3}[e-(23)][e+(13)]$, etc. Any such pair is satisfactory in view of their completeness together with S and A.

A neat procedure for dealing with these issues was devised by Young using boxes. For S_r we need r boxes:

r=2:

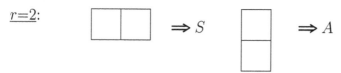

These are called **Young patterns**. When filled with labels they are called **Young tableaux**. Thus, corresponding to the symmetrizer S we have the tableaux

1	2

which says symmetrize in 1 and 2, i.e. apply the instruction $\frac{1}{2}[e+(12)]$, and corresponding to A we have the tableaux

1
2

which says antisymmetrize in 1 and 2, i.e. apply the instruction $\frac{1}{2}[e-(12)]$.
$\underline{r=3}$. There are now three patterns possible:

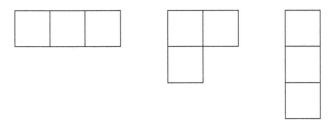

while the corresponding tableaux are:

$$\begin{array}{|c|c|c|}\hline 1 & 2 & 3 \\ \hline\end{array} = S = \frac{1}{6}\sum_R R$$

$$\begin{array}{|c|}\hline 1 \\ \hline 2 \\ \hline 3 \\ \hline\end{array} = A = \frac{1}{6}\sum_R \delta_R R$$

$$\begin{array}{|c|c|}\hline 1 & 2 \\ \hline 3 \\ \cline{1-1}\end{array} = Y = [e-(13)][e+(12)]/3$$

$$\begin{array}{|c|c|}\hline 1 & 3 \\ \hline 2 \\ \cline{1-1}\end{array} = Y' = [e-(12)][e+(13)]/3$$

$\underline{r=4}$. We now have five possible patterns and ten tableaux:

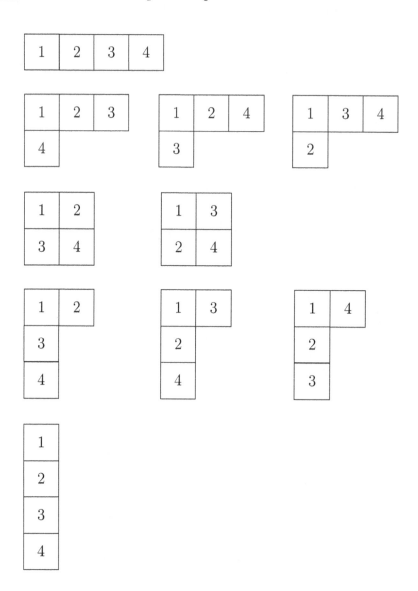

The rule for constructing different tableaux for a given pattern is as follows. The information in the tableaux can be written as MN, $M = M_1 M_2 M_3 \ldots$, where M_m instructs us to antisymmetrize in the entries in the mth column, and $N = N_1 N_2 N_3 \ldots$, where N_m instructs us to symmetrize in the entries of the mth row. We note that the M_i commute among themselves and the N_j commute among themselves but not with each other. Without loss of generality we agree to always place 1 in the upper left corner and to have entries increase from left to right in the rows and from top to bottom in the columns. Two tableaux are different if, subject to these rules, some of their entries are different.

The rule for constructing *patterns* is that they have a total of r boxes and every row be no longer than the row above it. Different patterns correspond to different irreducible representations of S_r, the number of different *tableaux* for a given pattern gives the dimension of that irreducible representation. The general pattern can be specified by the numbers $f_1 \geq f_2 \geq f_3 \geq ... \geq f_k$, where f_m denotes the number of boxes in the mth row and $f_1+f_2+f_3+...f_k=r$, i.e. we have a **partition** of r.

It is then obvious that for the pattern

there is only *one* tableaux possible obeying the rules:

1	2	3	...	r

and analogously only one tableaux is possible for the pattern consisting of a single column. These totally symmetric and totally antisymmetric representations are the only one-dimensional representations. The patterns with more than one column or row have larger dimensions, as we have seen above. As a check to determine whether we have found all possible tableaux we can use the fact that for a discrete group the sum of the squares of the dimensions of the irreducible representations should equal the dimension of the group. Thus:

S_2: $\qquad\qquad 1^2+1^2=2!$
S_3: $\qquad\qquad 1^2+2^2+1^2=3!$
S_4: $\qquad\qquad 1^2+3^2+2^2+3^2+1^2=4!$

The following formula gives the dimension of the irreducible representation of S_r specified by the partition $(f_1, f_2, ..., f_n)$:

$$\frac{r!}{\alpha_1!\alpha_2!...\alpha_n!} \prod_{1\leq j<k\leq n}(\alpha_j-\alpha_k), \qquad (14.29)$$

where α_k are the integers

$$\alpha_k = f_k+n-k, \quad k=1,2,...,n, \qquad (14.30)$$

in the decreasing order $\alpha_1 > \alpha_2 > ... > \alpha_n$.

Biographical Sketch

Young, Alfred (1873–1940) was born in Birchfield, England. He received the Sc.D. degree at Cambridge in 1908 and was ordained in the same year becoming curator at Christ Church, Blacklands. He wrote and published for over forty years over 25 papers on the subject of groups. Among them is a series of nine papers entitled "On quantitative substitutional analysis" in the first of which, published in 1900, he discusses the diagrams and tableaux that bear his name today.

15
Reduction of $SU(n)$ tensors

As discussed in preceding chapters the n^r entities $T_{\underbrace{abc...}_{r}}$ are the components of a contravariant $SU(n)$ tensor of rank r provided they have appropriate commutation relations with the generators of $SU(n)$. They can be used to provide representations of $SU(n)$, which are in general reducible. We get irreducible tensors if, under permutation, the subscripts belong to an irreducible representation of the symmetric group S_r, i.e. correspond to a Young pattern.

The tensors that are totally symmetric or totally antisymmetric under permutation of the subscripts are the easiest to consider. In particular, we shall now calculate their dimensions. Consider the totally antisymmetric tensor of rank r. The first index can take on n values, the second index can take on any value different from the first, i.e. $n-1$ values, and so on for the third index, etc. resulting in $n(n-1)...(n-r+1)$ possibilities. But exchanging any two indices simply gives a minus sign, so we must divide by $r!$ —thus the final answer for the number of independent components of a completely antisymmetric $SU(n)$ tensor of rank r is $n!/(n-r)!r!=\binom{n}{r}$, i.e. the number of ways of choosing r objects out of n. By a similar but slightly more involved argument one can show that the number of independent components of a completely symmetric $SU(n)$ tensor of rank r is $\binom{n+r-1}{r}$.

Next, to get a feeling for the subject, we find all the irreducible tensors and their dimensions for $r=2,3$ and 4.

r=2. A second rank $SU(n)$ tensor has n^2 components with the reduction

$$\square \times \square = \square\square + \begin{array}{c}\square\\\square\end{array}$$

$$n \times n = n(n+1)/2 + n(n-1)/2. \tag{15.1}$$

r=3. A third rank $SU(n)$ tensor has n^3 components with the reduction

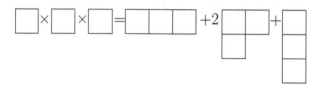

$$n \times n \times n = n(n+1)(n+2)/3! + 2 \times n(n^2-1)/3 + n(n-1)(n-2)/3!. \tag{15.2}$$

The best way to derive the above result is to use the results for $r=2$ and tack on the additional box in all possible ways as follows:

$$n \times n(n+1)/2 = n(n+1)(n+2)/3! + n(n^2-1)/3 \tag{15.3}$$

$$n \times n(n-1)/2 = n(n^2-1)/3 + n(n-1)(n-2)/3!. \tag{15.4}$$

r=4. We use the results for $r=3$ to obtain

$$n \times n(n+1)(n+2)/3! = (n+3)!/(n-1)!4! + 3n(n^2-1)(n+2)/4! \tag{15.5}$$

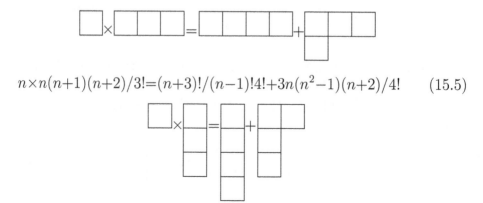

Reduction of SU(n) tensors

$$n \times n!/(n-3)!3! = n!/(n-4)!4! + 3n(n^2-1)(n-2)/4! \qquad (15.6)$$

$$n \times n(n^2-1)/3 = 3n(n^2-1)(n-2)/4! + 2n^2(n^2-1)/4!$$
$$+ 3n(n^2-1)(n+2)/4!. \qquad (15.7)$$

We are able to obtain the dimensions corresponding to a given pattern because in the sequence described at every stage there is at most one pattern that is not totally symmetric or totally antisymmetric or had appeared at an earlier stage of the sequence. This then allows us to obtain the dimension of that pattern.

It should be clear that what we have here is the Clebsch–Gordan series for the reduction of the Kronecker products of the defining representation with irreducible representations of rank one, two and three.

We should recognize some of these results. Let's start with $SU(2)$. We have $\square \to n \to 2$, i.e. the single box refers to the *defining two-dimensional* representation of $SU(2)$. In view of the isomorphism with $Spin(3)$ this is also our friend the spinor of $Spin(3)$. But now by tensoring this spinor representation with itself we get

$$2 \times 2 = 2 \cdot 3/2 + 2 \cdot 1/2 = 3 + 1, \qquad (15.8)$$

i.e. the totally symmetric representation is the triplet = spin 1, the totally antisymmetric representation is the singlet = spin 0 and we have the familiar statement that the addition of two spins 1/2 yields spin 1 and spin 0.

Next, we consider $r=3$:

$$2^3 = 2\cdot 3\cdot 4/3! + 2(2\cdot 3/3) + 2\cdot 1\cdot 0/3! = 4 + 2\cdot 2 + 0$$
$$= \text{spin}3/2 + 2(\text{spin}1/2). \tag{15.9}$$

What we are discovering is that the totally antisymmetric tensor of rank 3 vanishes in $SU(2)$ because it is not possible to have a structure with three slots (indices) completely antisymmetric under the exchange of slots if we only have *two* things to distribute among the *three* slots, the two things being the spin up and spin down of the $Spin(3)$ spinor.

Therefore, a Young pattern with a column having three or more boxes gives *zero* in $SU(2)$, which we have indicated above by shading such a column black. A column having two boxes transforms like a singlet, i.e. an invariant, and can be omitted if it is a part of a larger pattern—we have indicated this above by shading such a column grey. With this in mind we see that for $SU(2)$, for $r=4$, (15.6) is empty and (15.17) is equivalent to (15.2) so that the only non-trivial relation is (15.5):

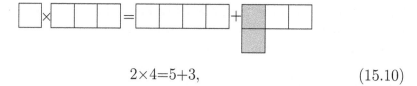

$$2\times 4 = 5 + 3, \tag{15.10}$$

that is addition of spin 1/2 and spin 3/2 yields spin 2 and spin 1.

In fact, it is obvious that for $SU(2)$ the only non-trivial arrangements correspond to the totally symmetric Young pattern consisting of a single row with p boxes of dimension $\binom{n+p-1}{p} \xrightarrow[n=2]{} p+1$, corresponding to spin $p/2$, where $p=0,1,2,...$

The situation is somewhat more interesting for $SU(3)$.

Now, $r=1$ means $\square \to n \to 3$, which is our friend the *quark*—the *defining three-dimensional* representation of $SU(3)$. By tensoring this we get for $r=2$:

$$3\times 3 = 6 + 3^*, \tag{15.11}$$

where the antisymmetric pattern is denoted as 3*, as will be explained below.

For $r=3$:

$$3\times 3\times 3 = 10 + 2\times 8 + 1. \qquad (15.12)$$

We now say a word about the notation 3*. In $SU(3)$ a column with three boxes is equivalent to a singlet (and may be omitted when part of a larger pattern), while a column with two boxes can be viewed as a missing third box, i.e. like the *absence* of the triplet = quark, i.e. like an *antiquark*. More precisely, a column with two boxes is like the conjugate of a single box according to the relation

$$\varepsilon^{abc} T_{bc} = S^a, \qquad (15.13)$$

which relates the antisymmetric rank two contravariant tensor T_{bc} to a rank one covariant tensor S^a. (This is analogous to the familiar fact that in 3-space the cross product of two vectors is a pseudovector—in 3-space a rank two antisymmetric tensor has the same number of components as a vector.)

In general, for $SU(n)$, the existence of the invariant tensor $\varepsilon^{\overbrace{ab...s}^{n}}$ means that if we identify a rank one contravariant tensor with a single box, a covariant rank one tensor can be related to a contravariant antisymmetric tensor of rank $n-1$ and identified with a column of $n-1$ boxes. When identifying these irreducible representations by their dimensions we shall use n and n^* to distinguish them. The existence of pairs of irreducible representations with the same dimension is a consequence of complex conjugation being an automorphism, with complex representations necessarily occurring as complex conjugate pairs.

It follows that the mixed rank two tensor corresponding to the adjoint representation is depicted as

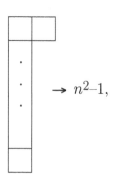 $\to n^2-1,$

(15.14)

where the long column in the above pattern has $n-1$ boxes.

It follows further that one can recognize from the pattern what representations are each other's conjugate and whether a representation is self-conjugate. Thus, for example, in $SU(4)$ the representations

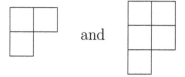

are each other's conjugate because if they are juxtaposed

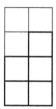

they form the pattern corresponding to the singlet. Similarly, it is easy to see that the adjoint representation is self-conjugate.

Whereas in $SU(2)$ the most general possible irreducible representation corresponds to a Young pattern with one row, in $SU(3)$ the most general case corresponds to two rows—and for $SU(n)$ to $n-1$ rows. This is why the most general irreducible representation of $SU(2)$ is specified by just *one* number—the integer $p=f_1-f_2$ corresponding to the number of columns with a single box, or to the spin $p/2$, or to the dimension $p+1$.

Correspondingly, in $SU(3)$ it takes *two* integers to uniquely label an irreducible representation: the integer $p=f_1-f_2$ corresponding to the number of columns with just one box, and $q=f_2-f_3$ corresponding to the number of columns with two boxes. This (p,q) way of labeling can also be thought of in terms of the number of quarks p and the number of antiquarks q, since a box $\square$ is like a quark, whereas $\begin{array}{c}\square\\\square\end{array}$ is like an antiquark.

Correspondingly, in $SU(n)$ we can specify the most general irreducible representation given by the partition $(f_1,f_2,...,f_n)$ by the $n-1$ integers $p_k=f_k-f_{k+1}, 1\leq k\leq n-1$. The following is a formula for the dimension of such an irreducible representation:

$$\dim(f_1,f_2,...,f_n) = \prod_{1\leq j<k\leq n}\left(1+\frac{f_j-f_k}{k-j}\right), \tag{15.15}$$

$$\xrightarrow[n=2]{} p_1+1$$

$$\xrightarrow[n=3]{} (p_1+1)(p_2+1)(p_1+p_2+2)/2$$

$$\xrightarrow[n=4]{} (p_1+1)(p_2+1)(p_3+1)\frac{p_1+p_2+2}{2}\frac{p_2+p_3+2}{2}\frac{p_1+p_2+p_3+3}{3}.$$

Reduction of the product of irreducible representations

Perhaps one of the most useful applications of the Young pattern techniques involves the reduction of the Kronecker product, i.e. the Clebsch–Gordan series, of irreducible representations. We have already used this to some extent—thus the relation (15.1)

$$\square \times \square = \square\square + \begin{array}{c}\square\\\square\end{array}$$

$$n\times n = n(n+1)/2 + n(n-1)/2$$

can be viewed as the statement that the reduction of the Kronecker product of the two defining n-dimensional representations yields the representations specified by the partitions $(2,0)$ and $(1,1)$.

The product of some arbitrary representation, say one specified by the partition $(4,2,2,1)$, with the defining n-dimensional representation reduces into the sum of the following:

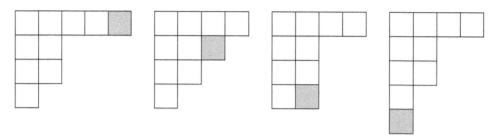

which is obtained by tacking on the extra (shaded) box to the pattern corresponding to (4,2,2,1) in all possible legal ways (meaning that in the resultant pattern every row is at least as long as the row below it). The resultant patterns are guaranteed to have the symmetry of the original pattern with respect to the boxes present in the original pattern corresponding to the partition (4,2,2,1); in addition the extra box has been incorporated and *no* symmetry clash occurs, because the extra box had *no* symmetry relation to any of the boxes in the (4,2,2,1) pattern.

However, when we combine two patterns, each of which has more than *one* box, we must worry about possible symmetry clash. We demonstrate the problem on two examples. The first is for $SU(2)$ where no clash occurs and we describe the most general reduction possible, the second is for $SU(n)$, $n>2$, where the problem is fully present.

As we know, the most general representation of $SU(2)$ can be specified by the spin j with the corresponding pattern consisting of a single row of $2j$ boxes. The most general reduction is therefore of a spin j_1 and a spin j_2 and we assume without loss of generality, that $j_1 \geq j_2$. We can form the pattern where the first row has $2j_1$ boxes, corresponding to the representation j_1, and the second row has $2j_2$ boxes, corresponding to the representation j_2. This involves no symmetry clash as the new pattern instructs us to first symmetrize the entries in each row—and that was true for each row to begin with—followed by antisymmetrizing the entries in each column, which again involves no conflict as the two entries in each column had no symmetry relation to begin with, coming from the two separate patterns. Since in $SU(2)$ a column with two boxes can be ignored this pattern is equivalent to a single row with $2(j_1-j_2)$ boxes, corresponding to spin j_1-j_2.

Next, we form the pattern where the first row has the $2j_1$ boxes of the representation j_1 and one additional box from the representation j_2, and the remaining $2j_2-1$ boxes of representation j_2 form the second row.

Again there is no symmetry clash. The resultant pattern is equivalent to a single row with $2(j_1-j_2+1)$ boxes, corresponding to spin j_1-j_2+1.

We can continue in this fashion, increasing the first row by one box at a time, until we reach the pattern consisting of a single row with $2j_1+2j_2$ boxes, corresponding to spin j_1+j_2. Thus, we arrive at the Clebsch–Gordan series

$$(2j_1+1)\times(2j_2+1)=\sum_{k=j_1-j_2}^{j_1+j_2}(2k+1), \tag{15.16}$$

which physicists should recognize as the well-known statement: the addition of two angular momenta j_1 and j_2 yields all angular momenta that range from $|j_1-j_2|$ to j_1+j_2.

We now look at $SU(n)$. The procedure for the reduction of the Kronecker product has been codified into a *set of rules*, which we will demonstrate on the example

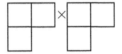

One starts out by taking one of the patterns to be the "skeleton" to which one tacks on the boxes from the other pattern (which pattern is chosen as the skeleton is immaterial, as a practical matter it is best to choose the one with more boxes). In the second pattern one labels all boxes in the first row "a", all boxes in the second row "b", all boxes in the third row "c", etc. Thus, in our example we have

One first tacks on to the skeleton the "a" boxes in all possible ways, except that two "a" boxes must *not* appear in the same column. Thus, in our example we obtain

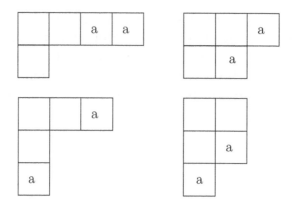

Next, one tacks on similarly the "b" boxes, then "c" boxes, etc. (no two "b" boxes, no two "c" boxes, etc., into the same column), but with the additional rule that, reading from right to left and top to bottom, the added-on boxes are in a lexical sequence. A sequence is **lexical** if it has the property that if we encounter in it at some point, say, three "c" next to each other, there were *before* at least three "b" (and therefore at least three "a"). Thus, we get in our example the following result:

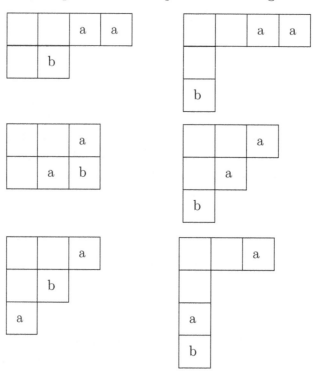

Reduction of SU(n) tensors

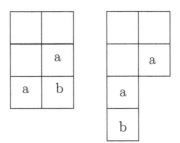

In summary then we have

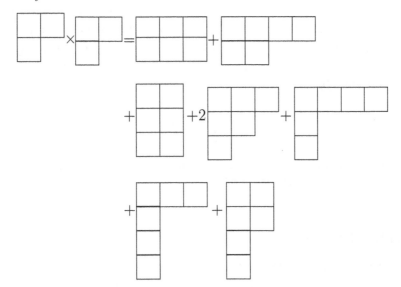

At the $SU(2)$ level this reads

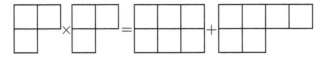

$$2 \times 2 = 1 + 3$$

while at the $SU(3)$ level this reads

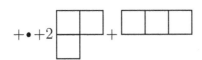

$$8\times 8 = 10^* + 27 + 1 + 2\cdot 8 + 10 \qquad (15.17)$$

(we use • to denote the pattern with n rows in $SU(n)$—the singlet).

We might note that since the **8** of $SU(3)$ is self-conjugate the representations contained in $\mathbf{8} \otimes \mathbf{8}$ must be either also self-conjugate (**1**, **8**, **27**) or must come in conjugate pairs (**10** and **10***). Another lesson that can be learned from the above is that a product of *three* **8**s in $SU(3)$ can be coupled to form an invariant (a singlet) in *two* different ways—these are the familiar F and D couplings of flavor theory where we couple Yukawa-like the antibaryon octet, the baryon octet and the meson octet.

I can't resist bringing up here the following problem: in how many different ways can one couple together r adjoint representations of $SU(n)$ to form a singlet? For $n \geq r$ the answer is given by the so-called number of derangements β_r

$$\beta_r = r! \sum_{k=0}^{r} \frac{(-1)^k}{k!}. \qquad (15.18)$$

This was apparently calculated first in 1708 by Montmort in "Essai d'Analyse sur les Jeux de Hasard".

Biographical Sketch

Kronecker, Leopold (1823–91) was born in Liegnitz, Germany (today Legnica, Poland). He studied under Kummer at Liegnitz Gymnasium and matriculated from the University of Berlin in 1841, then got his doctorate there in 1845 under Dirichlet. He wrote on number theory, elliptic functions, algebra and their interdependence. He became a full member of the Berlin Academy in 1861. In 1863 he succeeded Kummer in the chair of mathematics at the University of Berlin, becoming the first to hold the post who also earned the doctorate there. He opposed Kantor's views on transfinite numbers. He is famous for saying "God himself made whole numbers—everything else is the work of men."

16

Cartan basis, simple roots and fundamental weights

We recall that the generators X_K of a Lie algebra obey

$$[X_K, X_L] = C_{KL}{}^M X_M, \tag{16.1}$$

where the indices K, L, M range over d values, the dimension of the algebra. The Cartan metric is

$$g_{MN} = C_{MK}{}^L C_{NL}{}^K \tag{16.2}$$

and for a semisimple algebra it has an inverse g^{MN}

$$g^{MK} g_{KN} = \delta^M{}_N \tag{16.3}$$

and we can use g_{MN} to lower, and g^{MN} to raise, indices. The explicit form of the structure constants $C_{KL}{}^M$ depends on the choice of basis for the X_K. In particular, a certain basis was used by Cartan to arrive at the classification of all semisimple Lie algebras.

In preparation for the discussion of Cartan's classification of semisimple Lie algebras we now present the Cartan basis. In the process we will see that the basis used for the construction of $su(2)$ representations in Chapter 4 was a Cartan basis. In the Cartan basis the generators are grouped into two types. The generators called H_j are hermitian and mutually commute:

$$H_j^\dagger = H_j, \quad [H_j, H_k] = 0, \tag{16.4}$$

with lower case Latin indices ranging over r values, where r, the rank, is as large as possible. We recognize this r-dimensional Abelian algebra as the *Cartan subalgebra*.

The remaining $d-r$ generators are called E_α and satisfy

$$[H_j, E_\alpha] = \alpha_j E_\alpha, \tag{16.5}$$

i.e. E_α is an eigenvector of H_j to the eigenvalue α_j. Note that according to (16.4) H_k is also an eigenvector of H_j but to eigenvalue zero. We collect the α_j into an r-component vector $\boldsymbol{\alpha}$ called a **root**. Since the H_j are hermitian the α_j are real and the hermitian conjugate of (16.5) gives

$$[H_j,\ E_\alpha^\dagger] = -\alpha_j E_\alpha^\dagger \Rightarrow E_\alpha^\dagger = E_{-\alpha}, \tag{16.6}$$

that is the E_α are *not* hermitian and every root $\boldsymbol{\alpha}$ has associated to it $-\boldsymbol{\alpha}$. Note that therefore $d-r$ must be even.

Lastly, consider the commutator $[E_\alpha,\ E_\beta]$. From the Jacobi identity we get

$$[H_j,\ [E_\alpha,\ E_\beta]] = [E_\alpha,\ [H_j,\ E_\beta]] - [E_\beta,\ [H_j,\ E_\alpha]]$$
$$= (\alpha_j + \beta_j)[E_\alpha,\ E_\beta] \tag{16.7}$$

and we have three possibilities:

a) $\boldsymbol{\alpha} + \boldsymbol{\beta}$ is not a root therefore

$$[E_\alpha,\ E_\beta] = 0, \tag{16.8}$$

b) $\boldsymbol{\alpha}+\boldsymbol{\beta}$ is a zero root therefore

$$[E_\alpha,\ E_{-\alpha}] = N_\alpha{}^j H_j \tag{16.9}$$

c) $\boldsymbol{\alpha}+\boldsymbol{\beta}$ is a non-zero root therefore

$$[E_\alpha,\ E_\beta] = N_{\alpha\beta} E_{\alpha+\beta}. \tag{16.10}$$

Thus, the structure constants in this basis have been renamed

$$C_{ij}{}^M = 0,$$
$$C_{j\alpha}{}^M = \alpha_j \delta_\alpha{}^M,$$
$$C_{\alpha\beta}{}^M = N_{\alpha\beta} \delta_{\alpha+\beta}{}^M,\quad \boldsymbol{\alpha}+\boldsymbol{\beta} \neq 0,$$
$$C_{\alpha-\alpha}{}^m = N_\alpha{}^m,$$
$$C_{\alpha-\alpha}{}^\mu = 0. \tag{16.11}$$

Note that

$$N_\alpha{}^m = C_{\alpha-\alpha}{}^m = g^{mk} C_{\alpha-\alpha k} = g^{mk} C_{k\alpha-\alpha} = g_{-\alpha\beta} g^{mk} C_{k\alpha}{}^\beta$$
$$= g_{-\alpha\beta} g^{mk} \alpha_k \delta_\alpha{}^\beta = g_{-\alpha\alpha} \alpha^m. \tag{16.12}$$

Cartan basis, simple roots and fundamental weights

With E_α represented by matrices in the adjoint representation, we have

$$g_{-\alpha\alpha} = Tr E_\alpha^\dagger E_\alpha > 0 \qquad (16.13)$$

so we can rescale E_α and set

$$g_{-\alpha\alpha} = 1. \qquad (16.14)$$

For the Cartan metric we find

$$g_{ij} = \sum_\alpha \alpha_i \alpha_j \quad \Rightarrow \quad \sum_\alpha \alpha^k \alpha_k = r$$

$$g_{j\alpha} = 0$$

$$(1 - 2\alpha^k \alpha_k) g_{\alpha\beta} = N_{\alpha\nu} N_{-\alpha\alpha+\nu} \delta_{\alpha+\beta}{}^0. \qquad (16.15)$$

By looking at representations of the algebra we will show that $N_{\alpha\beta}$ can be expressed in terms of the roots. Therefore, all the information about the algebra is in the roots, which is how we shall find all possible semisimple Lie algebras.

We shall determine the $N_{\alpha\beta}$ as a special case from the $N_{\alpha\mu}$ where $\boldsymbol{\mu}$ is the **weight** vector, a vector in the r-dimensional root space with components μ_j. It labels states $|\mu\rangle$ in a representation of the algebra. So let us look at representations of the algebra.

Representations

Since the H_j are commuting hermitian operators they can be represented simultaneously by diagonal matrices

$$H_j |\mu\rangle = \mu_j |\mu\rangle, \qquad (16.16)$$

where $|\mu\rangle$ is an eigenstate of H_j to the eigenvalue μ_j. As is well known, eigenstates to different eigenvalues are orthogonal and without loss of generality we can normalize them:

$$\langle \nu | \mu \rangle = \delta_{\nu\mu}. \qquad (16.17)$$

The action of $E_{\pm\alpha}$ on these states is to raise or lower the weight $\boldsymbol{\mu}$ by $\boldsymbol{\alpha}$:

$$H_j E_{\pm\alpha} |\mu\rangle = ([H_j, E_{\pm\alpha}] + E_{\pm\alpha} H_j)|\mu\rangle$$
$$= (\mu \pm \alpha)_j E_{\pm\alpha} |\mu\rangle \qquad (16.18)$$

and therefore

$$E_{\pm\alpha}|\mu\rangle = N_{\pm\alpha\mu}|\mu\pm\alpha\rangle. \tag{16.19}$$

It should be clear that the $N_{\alpha\beta}$ introduced earlier are a special case of $N_{\alpha\mu}$ because *roots* are in fact *weights* in the adjoint representation. To find $N_{\pm\alpha\mu}$ we proceed as follows

$$\begin{aligned}\alpha^j\mu_j &= \langle\mu|\alpha^j H_j|\mu\rangle = \langle\mu|[E_\alpha, E_{-\alpha}]|\mu\rangle \\ &= \langle\mu|E^\dagger_{-\alpha} E_{-\alpha}|\mu\rangle - \langle\mu|E^\dagger_\alpha E_\alpha|\mu\rangle \\ &= |N_{-\alpha\mu}|^2 - |N_{\alpha\mu}|^2. \end{aligned} \tag{16.20}$$

From (16.19) we have

$$N_{-\alpha\mu} = \langle\mu-\alpha|E_{-\alpha}|\mu\rangle = \langle\mu|E_\alpha|\mu-\alpha\rangle^* = N_{\alpha\mu-\alpha}^* \tag{16.21}$$

so that

$$|N_{\alpha\mu-\alpha}|^2 - |N_{\alpha\mu}|^2 = \alpha^j\mu_j. \tag{16.22}$$

In a finite-dimensional representation this raising and lowering of μ by α must eventually end. Thus, after applying E_α to $|\mu\rangle$ say $(p+1)$ times we must get zero:

$$E_\alpha|\mu+p\alpha\rangle = |\mu+(p+1)\alpha\rangle N_{\alpha\mu+p\alpha} = 0 \tag{16.23}$$
$$\Rightarrow N_{\alpha\mu+p\alpha} = 0; \tag{16.24}$$

similarly, after applying $E_{-\alpha}$ to $|\mu\rangle$ say $(q+1)$ times we get zero:

$$N_{-\alpha\mu-q\alpha} = 0 = N_{\alpha\mu-(q+1)\alpha}. \tag{16.25}$$

Setting in (16.22) μ equal to $\mu+p\alpha$, then $\mu+(p-1)\alpha$, etc., results in the following chain (where we have introduced the notation $\alpha^j\mu_j \equiv \boldsymbol{\alpha}\cdot\boldsymbol{\mu}$):

$$|N_{\alpha\mu+(p-1)\alpha}|^2 - 0 = \alpha \cdot (\mu + p\alpha)$$
$$|N_{\alpha\mu+(p-2)\alpha}|^2 - |N_{\alpha\mu+(p-1)\alpha}|^2 = \alpha \cdot (\mu + (p-1)\alpha)$$

·

·

$$|N_{\alpha\mu}|^2 - |N_{\alpha\mu+\alpha}|^2 = \alpha \cdot (\mu + \alpha)$$
$$|N_{\alpha\mu-\alpha}|^2 - |N_{\alpha\mu}|^2 = \alpha \cdot \mu$$

·

·

$$|N_{\alpha\mu-q\alpha}|^2 - |N_{\alpha\mu-(q-1)\alpha}|^2 = \alpha \cdot (\mu - (q-1)\alpha)$$
$$0 - |N_{\alpha\mu-q\alpha}|^2 = \alpha \cdot (\mu - q\alpha). \tag{16.26}$$

Adding all these equations yields

$$0 = \sum_{k=-q}^{p} \alpha \cdot (\mu + k\alpha) = (p+q+1)(\alpha \cdot \mu + \alpha \cdot \alpha \tfrac{p-q}{2}). \tag{16.27}$$

Since p and q are, by construction non-negative integers we arrive at the important result

$$2\alpha \cdot \mu / \alpha \cdot \alpha = q - p = \text{integer}, \tag{16.28}$$

while the recursion for the N yields

$$|N_{\alpha\mu}|^2 = p\alpha \cdot (\mu + \tfrac{p+1}{2}\alpha) = p(q+1)\alpha \cdot \alpha / 2. \tag{16.29}$$

It follows from (16.29), among other things, that

$$\alpha \cdot \alpha = g_{ij} \alpha^i \alpha^j > 0 \tag{16.30}$$

that is the Cartan metric restricted to the Cartan subalgebra is *positive definite* and the r-dimensional vector space of the roots and weights is *Euclidean*, therefore without loss of generality we can take

$$g_{ij} = \delta_{ij} \tag{16.31}$$

and no distinction need be made between covariant and contravariant components of root and weight vectors.

Equation (16.28) has the following important consequence. We have the string of weights

$$\mu+k\alpha, \quad -q\leq k\leq p \tag{16.32}$$

and $\mu-(q+1)\alpha$, $\mu+(p+1)\alpha$ are not weights. For a weight ν we define $(\nu)_{W\alpha}$ to be the reflection of ν through a hyperplane orthogonal to the root α:

$$(\nu)_{W\alpha}=\nu-2\alpha\nu\cdot\alpha/\alpha\cdot\alpha. \tag{16.33}$$

Then,

$$(\mu+k\alpha)_{W\alpha}=\mu+k\alpha-2\alpha\cdot(\mu+k\alpha)/\alpha\cdot\alpha$$
$$=\mu+(p-q-k)\alpha \tag{16.34}$$

and as k takes values between $-q$ and p so does $(p-q-k)$. Thus, we have the important result: if ν is a weight and α is a root then $(\nu)_{W\alpha}$ is also a weight. The hyperplanes orthogonal to the roots are called *Weyl hyperplanes*, the reflections in these planes are called *Weyl reflections*, and the totality of these reflections and their products form a group called the **Weyl group**.

All of our results about weights are also true if we replace everywhere a weight by a root because, as noted before, roots are weights in the adjoint representation. In particular, we have the analog of (16.28)

$$2\alpha\cdot\beta/\alpha\cdot\alpha=q-p=\text{integer}. \tag{16.35}$$

Next, we introduce an *ordering* of the weights. We call a weight **positive** if its first non-zero component starting from the left is positive, we call a weight negative if its first non-zero component is negative and we call it zero if all the components are zero. Obviously every weight is either positive, negative or zero and this definition provides an (arbitrary) splitting of the space of non-zero weights in two.

Then, weights can be compared:

$$\mu>\nu \tag{16.36}$$

if $\mu-\nu$ is positive. Therefore, in a representation we must have a state of *highest weight*, whose weight vector will be denoted by **h**. If then we use φ to denote some *positive* root we can conclude from our master formula that

Cartan basis, simple roots and fundamental weights **135**

$$2\mathbf{h}\cdot\boldsymbol{\varphi}/\boldsymbol{\varphi}\cdot\boldsymbol{\varphi}=q=\text{non-negative integer} \tag{16.37}$$

because $p=0$, because $\mathbf{h}+\boldsymbol{\varphi}$ is not a weight, because $\mathbf{h}$ is highest and $\boldsymbol{\varphi}$ is positive.

Even better, we have

$$2\mathbf{h}\cdot\boldsymbol{\alpha}^{(i)}/\boldsymbol{\alpha}^{(i)}\cdot\boldsymbol{\alpha}^{(i)}=q^i=\text{non-negative integer,} \tag{16.38}$$

(*no* sum over i) where the $\boldsymbol{\alpha}^{(i)}$ are simple roots, which we now proceed to define.

We would like to find a set of r roots that could serve as a basis in the r-dimensional space. We know that the number of roots is $d-r$, which is generally much larger than r. We know that we can split the space of roots in two: a half that are positive and a half that are negative—but even $(d-r)/2$, the number of positive roots, is generally larger than r.

We define a **simple root** as a positive root that cannot be expressed as an arithmetic sum of two positive roots. What we mean is this: suppose φ_1 and φ_2 are positive roots and φ is a root such that $\varphi=\varphi_1+\varphi_2$. Then φ is a positive root but *not* a simple root. On the other hand, if $\boldsymbol{\alpha}$ is a simple root then $\boldsymbol{\alpha}\neq\varphi_1+\varphi_2$.

It then follows after a little thought that

1) Every positive root φ is expressible in the form

$$\varphi=\sum_{s=1}^{r}k_s\boldsymbol{\alpha}^{(s)}, \tag{16.39}$$

where the $\boldsymbol{\alpha}^{(s)}$ are simple roots and the k_s are *non-negative integers*. Recall that the roots are r-dimensional vectors and we use the superscript to distinguish different vectors (thus $\alpha_j^{(s)}$ is the jth component of the sth simple root).

It is an immediate corollary that every *negative* root is similarly expressible in terms of simple roots with coefficients that are *non-positive* integers.

2) An expression of the form

$$\sum_{s=1}^{r} x_s \boldsymbol{\alpha}^{(s)} \tag{16.40}$$

is *not* a root except if *all* the x_s are non-negative integers or non-positive integers. Thus,

$$\alpha^{(1)}+\alpha^{(2)}/2 \text{ is not a root,}$$
$$\alpha^{(1)}+\sqrt{3}\alpha^{(2)} \text{ is not a root,}$$
$$\alpha^{(1)}+\alpha^{(2)}-\alpha^{(3)} \text{ is not a root, etc.}$$

3) Lastly, we comment on the fact that the number of different simple roots is r. Clearly we want them to be linearly independent, hence their number cannot exceed r. But suppose that their number were less than r. Then we could arrange our basis so that, say, the first component of all α vanished meaning

$$[H_1, E_\varphi]=0 \text{ for all roots } \varphi. \tag{16.41}$$

Now of course $[H_1, H_j]=0$ for all j and so we find that H_1 commutes with all the generators and forms an invariant (Abelian) subalgebra in violation of the requirement of the algebra being semisimple. Thus, the simple roots form a basis.

We now go back and note that the r integers q^i defined by (16.38) carry the same information as the highest weight because $\boldsymbol{h}$ is an r-dimensional vector and the q^i specify the projection of $2\boldsymbol{h}$ onto the $\boldsymbol{\alpha}^{(i)}$ that are a basis in this r-dimensional space.

To get a better understanding of what this means define a set of weight vectors $\boldsymbol{\mu}^{(j)}$, $1 \le j \le r$, by the relation

$$2\boldsymbol{\mu}^{(j)} \cdot \boldsymbol{\alpha}^{(i)} / \boldsymbol{\alpha}^{(i)} \cdot \boldsymbol{\alpha}^{(i)} = \delta^{ji}, \tag{16.42}$$

i.e. $\boldsymbol{\mu}^{(j)}$ is the *highest weight* of some representation for which all q^i are zero except when $i=j$ for which $q^j=1$. These representations, r in number, are called **fundamental representation** and the $\boldsymbol{\mu}^{(j)}$ are called **fundamental weight**. The relation (16.42) suggests that the fundamental weights are *dual* to simple roots and just as the simple roots span the space of roots the fundamental weights span the space of highest weights. Indeed, it follows from (16.38) and (16.42) that any highest weight $\boldsymbol{h}$ can be written in terms of the fundamental weights as

$$\boldsymbol{h} = q^j \boldsymbol{\mu}^{(j)}. \tag{16.43}$$

In words: any highest weight can be expressed as the sum of fundamental weights with non-negative integer coefficients—and this statement is the dual of: any positive root can be expressed as the sum of simple roots with non-negative integer coefficients. There are similar dual statements about other properties of simple roots and fundamental weights.

Equation (16.43) suggests an algorithm for obtaining all possible irreducible representations for an algebra for which a set of simple roots is known by forming all possible h as follows: find all the fundamental weights using (16.42). Form the representation F^j by assigning to it as its highest weight the weight $\boldsymbol{\mu}^{(j)}$. Then, construct the tensor product

$$\underbrace{F^1 \otimes F^1 ... \otimes F^1}_{q^1} \otimes \underbrace{F^2 \otimes F^2 ... \otimes F^2}_{q^2} \otimes ... \underbrace{F^r \otimes ... \otimes F^r}_{q^r} \qquad (16.44)$$

—the state of highest weight in that product will have weight $\boldsymbol{h} = q^j \boldsymbol{\mu}^{(j)}$ as desired.

Throughout we have been assuming that the representation is determined by the state of highest weight and that indeed is the case as all the states in the representation are, by definition, the states obtained from the highest weight state by repeated application of generators corresponding to all possible negative roots.

Application to $su(2)$

We illustrate all of this formalism on the example of $su(2)$. We recall (4.5) from Chapter 4

$$[J_3, J_\pm] = \pm J_\pm, \quad [J_+, J_-] = 2J_3. \qquad (16.45)$$

The Cartan subalgebra consists of the single generator J_3, so the rank is one. We have a single simple root with a single component $\alpha_3 = 1$ and a single negative root $-\alpha_3 = -1$. These two roots are Weyl reflections of each other across a line orthogonal to them. The non-zero structure constants and the Cartan metric are

$$C_{3\pm}{}^\pm = \pm 1, \quad C_{+-}{}^3 = 2, \quad g_{33} = 2, \quad g_{+-} = g_{-+} = 4. \qquad (16.46)$$

The inverse of g_{33} is $g^{33} = 1/2$ so that

$$\alpha^3 \alpha_3 = g^{33} \alpha_3 \alpha_3 = 1/2 \qquad (16.47)$$

and we have the expected relation

$$\alpha^3 \alpha_3 + (-\alpha^3)(-\alpha_3) = 1 = \text{rank}. \qquad (16.48)$$

There is a single fundamental representation with highest weight μ_3 and it follows from the definition

$$2\mu_3\alpha^3/\alpha_3\alpha^3=1 \tag{16.49}$$

that

$$\mu_3=\alpha_3/2=1/2. \tag{16.50}$$

The "one" on the right-hand-side of (16.49) means that we can subtract α_3 from μ_3 once, but not twice, and get a valid weight. Since α_3 is the only simple root the fundamental representation is two-dimensional, consisting of the two states $|1/2\rangle$ and $J_-|1/2\rangle \sim |-1/2\rangle$.

To get an arbitrary highest weight h, here called j, we take the highest weight in the n-fold tensoring of the fundamental representation:

$$j=n/2, \quad n=0,1,2,3,.... \tag{16.51}$$

Applying $(J_-)^k$ to this state $|j\rangle$ produces a state of weight $j-k$. Because applying J_- to the highest weight state in the fundamental representation more than once gives zero it follows that in the representation with highest weight j applying J_- $n+1$ times produces zero, i.e. the state of lowest weight has weight $j-n=-j$ and the dimension of the representation is $2j+1$. In particular, the adjoint representation corresponds to $j=1$.

Explicitly, denoting by $|m\rangle$ the states in the representation specified by j, we have

$$J_3|m\rangle=m|m\rangle, \quad J_\pm|m\rangle=N_{\pm\alpha_3 m}|m\pm 1\rangle$$
$$|N_{\alpha_3 m}|^2=p(q+1)\alpha^3\alpha_3/2=(j-m)(j+m+1) \tag{16.52}$$

since

$$m+p\alpha^3=j, \quad m-(q+1)\alpha^3=-(j+1) \tag{16.53}$$

and we recognize all of the results of Chapter 4.

Lastly, we note that by rescaling the generators $J_3 \to J_3'=J_3/\sqrt{2}, J_\pm \to J_\pm'=J_\pm/2$ we can make all non-zero elements of the Cartan metric equal to unity (and $\alpha_3'=\alpha'^3=1/\sqrt{2}$).

17
Cartan classification of semisimple algebras

We look again at the master formula

$$2\mu\cdot\alpha/\alpha\cdot\alpha = q-p, \tag{17.1}$$

where q is the number of times we can subtract α from μ and still get a weight, p is the number of times we can add α to μ and still get a weight. Setting $\mu=\beta$, a root, we have

$$2\beta\cdot\alpha/\alpha\cdot\alpha = m. \tag{17.2}$$

Of course we also have

$$2\alpha\cdot\beta/\beta\cdot\beta = n, \tag{17.3}$$

where m and n are integers, positive, negative or zero.

Therefore,

$$n/m = \alpha\cdot\alpha/\beta\cdot\beta \tag{17.4}$$

and

$$nm = 4(\alpha\cdot\beta)^2/\alpha\cdot\alpha\beta\cdot\beta = 4\cos^2\vartheta, \tag{17.5}$$

where ϑ is the angle between α and β. Here, we use the fact that our r-dimensional space is Euclidean so the dot product can be expressed in terms of cosines.

Since $0 \leq \cos^2\vartheta \leq 1$ it follows that the number of allowed values for m and n, and therefore for ϑ, is quite limited. We can make the following table.

Table 17.1 Allowed angles

m	n	$\cos\vartheta$	ϑ
0	0	0	90°
1	1	$\sqrt{\frac{1}{4}}$	60°
−1	−1	$-\sqrt{\frac{1}{4}}$	120°
2	1	$\sqrt{\frac{2}{4}}$	45°
−2	−1	$-\sqrt{\frac{2}{4}}$	135°
3	1	$\sqrt{\frac{3}{4}}$	30°
−3	−1	$-\sqrt{\frac{3}{4}}$	150°
2	2	1	0°
−2	−2	−1	180°

We rewrite (17.5) to incorporate these results:

$$\alpha\cdot\beta = \pm\tfrac{1}{2}\sqrt{n}\;\alpha\beta, \quad n=0,1,2,3,4, \qquad (17.6)$$

(where we write α^2 for $\alpha\cdot\alpha$, etc.) and if σ and τ stand for *distinct simple* roots we have the stronger statement

$$\sigma\cdot\tau = -\tfrac{1}{2}\sqrt{n}\;\sigma\tau, \quad n=0,1,2,3. \qquad (17.7)$$

The dot product for simple roots being negative is due to the q in the corresponding (17.1) being zero. If q were not zero $\rho=\sigma-\tau$ would be a root, positive or negative; if positive then we can write $\sigma=\rho+\tau$, if negative then we can write $\tau=\sigma+(-\rho)$, in either case we have a simple root given as a sum of positive roots in violation of the definition of simple roots.

These remarkable results mean that we can draw in our r-dimensional Euclidean space a *root diagram* (= weight diagram for the adjoint representation) in a very limited number of ways leading to a complete classification of all possible semisimple Lie algebras.

Rank 1

The root space is *one-dimensional*, we have one generator H in the Cartan subalgebra, one raising generator E_-, one lowering generator $E_-=E_+^\dagger$, one simple root σ_+. The commutation relations are

$$[H, E_+] = \tfrac{1}{\sqrt{2}} E_+, \quad [H, E_-] = -\tfrac{1}{\sqrt{2}} E_-, \quad [E_+, E_-] = H, \qquad (17.8)$$

where the root $\sigma_+ = \tfrac{1}{\sqrt{2}}$ because [see (16.15)]

$$\sigma_+ \sigma_+ + \sigma_- \sigma_- = \text{rank} = 1 \quad \text{and} \quad \sigma_- = -\sigma_+. \qquad (17.9)$$

The root diagram that contains all this information is

weight	$-\tfrac{1}{\sqrt{2}}$	0	$\tfrac{1}{\sqrt{2}}$
generator	E_-	H	E_+

We remark that any root α provides such a triplet. To see that put in the master formula (17.1) $\mu = \alpha$ and find

$$2 = q - p. \qquad (17.10)$$

Now, p must be zero since if it were not then 2α would be a root but

$$E_{2\alpha} \sim [E_\alpha, E_\alpha] = 0, \qquad (17.11)$$

therefore $q = 2$ and

$$\alpha, \quad \alpha - \alpha = 0, \quad \alpha - 2\alpha = -\alpha \qquad (17.12)$$

is a complete chain of roots.

In Cartan's classification this algebra is $A_1 = B_1 = C_1$ and we recognize it as $su(2) \cong so(3) \cong sp(2)$.

Rank 2

We have two simple roots that we shall call σ and τ. For distinct simple roots (17.1) becomes

$$2\sigma \cdot \tau / \tau^2 = -p, \quad 2\tau \cdot \sigma / \sigma^2 = -p', \quad p, p' \text{ positive integers.} \qquad (17.13)$$

a) We start with the case when the angle between the two simple roots is 90° and $pp' = 0$ so that the sum of the roots is not a root. Thus, we have an algebra of dimension six: the two simple roots, their negatives, and two zero roots. The root diagram is simply two rank one diagrams at right angles to each other. There is no connection between the two simple roots, so in fact the diagram can be separated into two disjoint parts. This algebra is called D_2 in Cartan's classification and we have $D_2 = A_1 \oplus A_1$—we recognize $so(4)$.

b) Next, let the angle between the simple roots be 120° and $p=p'=1$. That means that we have the roots

$$\pm\sigma, \quad \pm\tau, \quad \pm(\sigma+\tau), \tag{17.14}$$

all of the same length and the dimension of the algebra is eight.

From (16.43) we have for the fundamental weight $\boldsymbol{\mu}^{(1)}$ the definition

$$2\boldsymbol{\mu}^{(1)}\cdot\boldsymbol{\sigma}/\sigma^2 = 1, \quad 2\boldsymbol{\mu}^{(1)}\cdot\boldsymbol{\tau}/\tau^2 = 0 \tag{17.15}$$

with the solution

$$\boldsymbol{\mu}^{(1)} = (2\boldsymbol{\sigma}+\boldsymbol{\tau})/3. \tag{17.16}$$

Other weights in this fundamental representation are $\boldsymbol{\mu}^{(1)} - \boldsymbol{\sigma} = (\boldsymbol{\tau}-\boldsymbol{\sigma})/3$ and $\boldsymbol{\mu}^{(1)} - \boldsymbol{\sigma} - \boldsymbol{\tau} = -(\boldsymbol{\sigma}+2\boldsymbol{\tau})/3$, thus this representation is 3-dimensional.

The fundamental weight $\boldsymbol{\mu}^{(2)}$ and the corresponding representation are obtained by interchanging $\boldsymbol{\sigma}$ and $\boldsymbol{\tau}$. If we recall that $su(3)$ is of rank two and dimension eight, and that its defining representation (as well as its conjugate) is 3-dimensional we see that this algebra, which Cartan calls A_2, is $su(3)$.

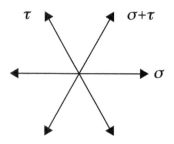

Fig. 17.1 Root diagram for A_2

c) Next, let the angle between the simple roots be 135° and $p=1$, $p'=2$. Let σ be the short root and τ the long root: $\tau^2 = 2\sigma^2$. That means that we have the roots

$$\pm\sigma, \quad \pm\tau, \quad \pm(\sigma+\tau), \quad \pm(2\sigma+\tau) \tag{17.17}$$

and this algebra is ten-dimensional. The fundamental weights are

$$\mu^{(1)}=\sigma+\tau, \quad \mu^{(2)}=\sigma+\tau/2, \tag{17.18}$$

the representation corresponding to $\mu^{(1)}$ is 5-dimensional, that corresponding to $\mu^{(2)}$ is 4-dimensional. If we recall that $so(5)$ is of rank two and dimension ten, that its defining representation is 5-dimensional and its spinor representation is 4-dimensional we see that this algebra, which Cartan calls B_2, is $so(5)$. It is also C_2, which we call $sp(4)$.

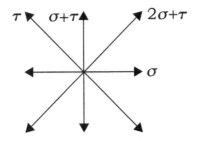

Fig. 17.2 *Root diagram for $B_2=C_2$*

d) Lastly, let the angle be 150° and $p=1$, $p'=3$. Let σ be the short root and τ the long root: $\tau^2=3\sigma^2$. We have six short roots: $\pm\sigma$, $\pm(\sigma+\tau)$, $\pm(2\sigma+\tau)$, and six long roots: $\pm\tau$, $\pm(3\sigma+\tau)$, $\pm(3\sigma+2\tau)$ and this algebra is 14-dimensional. The fundamental weights are

$$\mu^{(1)}=2\sigma+\tau, \quad \mu^{(2)}=3\sigma+2\tau. \tag{17.19}$$

The representation corresponding to $\mu^{(1)}$ is 7-dimensional, that corresponding to $\mu^{(2)}$ is 14-dimensional. Cartan called this algebra G_2, it does not correspond to any of the classical algebras and is one of the so-called exceptional algebras.

Adding the lengths squared of all roots gives 2 (the rank), therefore the length squared of the D_2 roots is 1/2, of the A_2 roots is 1/3, of the short B_2 roots is 1/6 and of the short G_2 roots is 1/12.

Rank n

We have been using as a basis in root space the simple roots, which are, in general, not orthogonal. For the purpose of generalizing from rank two to rank n it turns out to be convenient to introduce an orthogonal basis.

a) The $A_n=su(n+1)$ series.

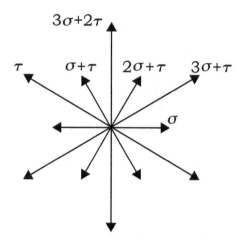

Fig. 17.3 Root diagram for G_2

Although the root space of A_n is n-dimensional the description in terms of $n+1$ dimensions is more transparent. Essentially this is because $u(n+1)$ is of rank $n+1$ and dealing with $u(n+1)$, and then restricting it to $su(n+1)$, is more transparent than dealing with $su(n+1)$ directly.

As a first step we rewrite the A_2 results in terms of $\mathbf{e}_1$, $\mathbf{e}_2$ and $\mathbf{e}_3$, a set of orthogonal vectors such that

$$\mathbf{e}_j \cdot \mathbf{e}_k = \delta_{jk}/6, \quad 1 \leq j,k \leq 3, \tag{17.20}$$

in terms of which the simple roots of A_2 are given by

$$\sigma = \mathbf{e}_1 - \mathbf{e}_2, \quad \tau = \mathbf{e}_2 - \mathbf{e}_3. \tag{17.21}$$

It follows that all the roots of A_2 are given by

$$\mathbf{e}_j - \mathbf{e}_k, \quad 1 \leq j \neq k \leq 3 \tag{17.22}$$

and the simple roots can be restated as

$$\mathbf{e}_j - \mathbf{e}_{j+1}, \quad 1 \leq j \leq 2. \tag{17.23}$$

The two fundamental weights are

$$\boldsymbol{\mu}^{(1)} = \mathbf{e}_1 - \tfrac{1}{3}(\mathbf{e}_1 + \mathbf{e}_2 + \mathbf{e}_3), \tag{17.24}$$
$$\boldsymbol{\mu}^{(2)} = \mathbf{e}_1 + \mathbf{e}_2 - \tfrac{2}{3}(\mathbf{e}_1 + \mathbf{e}_2 + \mathbf{e}_3),$$

which can be restated as

$$\mu^{(j)} = w_j - \tfrac{j}{3} w_3, \quad 1 \le j \le 2, \tag{17.25}$$

where

$$w_s = \sum_{k=1}^{s} e_k, \quad 1 \le s \le 3. \tag{17.26}$$

Although the e_j, $1 \le j \le 3$, span a 3-dimensional space, all the roots and weights above live in the 2-dimensional space orthogonal to $w_3 = e_1 + e_2 + e_3$.

This suggests that we define the A_n algebra by the following generalization

$$e_j \cdot e_k = \delta_{jk}/2(n+1), \quad 1 \le j,k \le n+1, \tag{17.27}$$

all roots: $\quad e_j - e_k, \quad 1 \le j \ne k \le n+1, \tag{17.28}$

simple roots: $\quad e_j - e_{j+1}, \quad 1 \le j \le n, \tag{17.29}$

fundamental weights: $\quad \mu^{(j)} = w_j - \tfrac{j}{n+1} w_{n+1}, \quad 1 \le j \le n, \tag{17.30}$

$$w_s = \sum_{k=1}^{s} e_k, \quad 1 \le s \le n+1, \tag{17.31}$$

and while the e_j, $1 \le j \le n+1$, span an $(n+1)$-dimensional space the roots and weights live in the n-dimensional space orthogonal to w_{n+1}. It is straightforward but tedious to show that the roots (17.28) obey (17.6) and are closed under the Weyl group.

b) The $B_n = so(2n+1)$ series.

The $B_2 = C_2$ can be generalized in two different ways, giving rise to the B_n and C_n series that are different for n larger than two. To obtain the B_n series we rewrite the $B_2 = C_2$ results in terms of two orthogonal vectors e_1 and e_2 that satisfy

$$e_j \cdot e_k = \delta_{jk}/6, \quad 1 \le j,k \le 2, \tag{17.32}$$

in terms of which the simple roots of $B_2 = C_2$ are given by

$$\sigma = e_2, \quad \tau = e_1 - e_2, \tag{17.33}$$

all the roots of $B_2=C_2$ are given by

$$e_j-e_k, \quad \pm(e_j+e_k), \quad 1\leq j\neq k\leq 2, \quad \pm e_j, \quad 1\leq j\leq 2, \qquad (17.34)$$

and the fundamental weights of $B_2=C_2$ are given by

$$\mu^{(1)}=e_1, \quad \mu^{(2)}=(e_1+e_2)/2. \qquad (17.35)$$

Guided by these expressions we generalize to obtain the B_n series of algebras as follows:

$$e_j\cdot e_k=\delta_{jk}/2(2n-1), \quad 1\leq j,k\leq n \qquad (17.36)$$

simple roots: $\quad e_n, \quad e_j-e_{j+1}, \quad 1\leq j\leq n-1, \qquad (17.37)$

all roots: $\quad e_j-e_k, \quad \pm(e_j+e_k), \quad 1\leq j\neq k\leq n,$

$$\pm e_j, \quad 1\leq j\leq n, \qquad (17.38)$$

fundamental weights: $\quad \mu^{(j)}=w_j, \quad 1\leq j\leq n-1, \quad \mu^{(n)}=w_n/2. \qquad (17.39)$

c) The $C_n=sp(2n)$ series.

The other way to generalize $B_2=C_2$ is to introduce e_1 and e_2 that satisfy

$$e_j\cdot e_k=\delta_{jk}/12, \quad 1\leq j,k\leq 2, \qquad (17.40)$$

and write the $B_2=C_2$ simple roots as

$$\sigma=e_1-e_2, \quad \tau=2e_2. \qquad (17.41)$$

In terms of the root diagram, Fig. 17.2, the present assignment amounts to a rotation of the diagram by 45°.

All the roots of $B_2=C_2$ are now given by

$$e_j-e_k, \quad 1\leq j\neq k\leq 2,$$
$$\pm(e_j+e_k), \quad 1\leq j\neq k\leq 2, \qquad (17.42)$$
$$\pm 2e_j, \quad 1\leq j\leq 2,$$

and the fundamental weights of $B_2=C_2$ are now given by

$$\mu^{(1)}=e_1, \quad \mu^{(2)}=(e_1+e_2), \qquad (17.43)$$

where we have exchanged the superscripts (1) and (2).

Cartan classification of semisimple algebras 147

We now generalize these expressions to obtain the C_n series:
$$e_j \cdot e_k = \delta_{jk}/4(n+1), \quad 1 \leq j,k \leq n, \tag{17.44}$$

simple roots: $\quad 2e_n, \quad e_j - e_{j+1}, \quad 1 \leq j \leq n-1, \tag{17.45}$

all roots: $\quad e_j - e_k, \quad \pm(e_j + e_k), \quad 1 \leq j \neq k \leq n,$
$$\pm e_j, \quad 1 \leq j \leq n, \tag{17.46}$$

fundamental weights: $\mu^{(j)} = w_j, \quad 1 \leq j \leq n-1, \quad \mu^{(n)} = w_n. \tag{17.47}$

d) The $D_n = so(2n)$ series.

We introduce e_1 and e_2 satisfying
$$e_j \cdot e_k = \delta_{jk}/4, \quad 1 \leq j,k \leq 2, \tag{17.48}$$

in terms of which the simple roots are given by
$$e_1 + e_2, \quad e_1 - e_2, \tag{17.49}$$

all roots are given by
$$e_j - e_k, \quad 1 \leq j,k \leq 2,$$
$$\pm(e_j + e_k), \quad 1 \leq j \neq k \leq 2, \tag{17.50}$$

and fundamental weights are given by
$$\mu^{(\pm)} = (e_1 \pm e_2)/2. \tag{17.51}$$

We now generalize these expressions to obtain the D_n series:
$$e_j \cdot e_k = \delta_{jk}/4(n-1), \quad 1 \leq j,k \leq n, \tag{17.52}$$

simple roots: $\quad e_{n-1} + e_n, \quad e_j - e_{j+1}, \quad 1 \leq j \leq n-1 \tag{17.53}$

all roots: $\quad e_j - e_k, \quad 1 \leq j,k \leq n,$
$$\pm(e_j + e_k), \quad 1 \leq j \neq k \leq n, \tag{17.54}$$

fundamental weights: $\mu^{(j)} = w_j, \quad 1 \leq j \leq n-2,$
$$\mu^{(\pm)} = (w_{n-1} \pm e_n)/2. \tag{17.55}$$

e) Generalization of G_2.

Guided by the A_n, B_n, C_n and D_n cases we attempt to generalize G_2. We recall that the two simple roots of G_2 satisfy

$$\tau^2 = 3\sigma^2 = 1/4, \quad 2\boldsymbol{\sigma}\cdot\boldsymbol{\tau} = -3\sigma^2. \tag{17.56}$$

We attempt to obtain G_3 by adding a simple root $\boldsymbol{\rho}$ in the form

$$\boldsymbol{\rho} = a\boldsymbol{\sigma} + b\boldsymbol{\tau} + \boldsymbol{c}, \tag{17.57}$$

where a, b and $\boldsymbol{c}$ are arbitrary except that $\boldsymbol{c}$ is non-zero and orthogonal to $\boldsymbol{\sigma}$ and $\boldsymbol{\tau}$ to ensure that $\boldsymbol{\rho}$, $\boldsymbol{\sigma}$ and $\boldsymbol{\tau}$ are linearly independent:

$$\boldsymbol{c}\cdot\boldsymbol{\sigma} = \boldsymbol{c}\cdot\boldsymbol{\tau} = 0, \quad \boldsymbol{c} \neq 0. \tag{17.58}$$

We have from (17.7)

$$\boldsymbol{\rho}\cdot\boldsymbol{\sigma} = a\sigma^2 + b\boldsymbol{\sigma}\cdot\boldsymbol{\tau} = (a - 3b/2)\sigma^2 = -\tfrac{1}{2}\sqrt{n_1}\,\rho\sigma,$$

$$\boldsymbol{\rho}\cdot\boldsymbol{\tau} = a\boldsymbol{\sigma}\cdot\boldsymbol{\tau} + b\tau^2 = 3(b - a/2)\sigma^2 = -\tfrac{1}{2}\sqrt{n_2}\,\rho\tau, \tag{17.59}$$

so that

$$a\sigma = -(2\sqrt{n_1} + \sqrt{3n_2})\rho$$

$$b\sigma = -(\sqrt{n_1} + 2\sqrt{n_2/3})\rho. \tag{17.60}$$

Therefore,

$$\rho^2 = \boldsymbol{\rho}\cdot(a\boldsymbol{\sigma} + b\boldsymbol{\tau} + \boldsymbol{c}) = -\tfrac{1}{2}\sqrt{n_1}\,\rho\sigma a - \tfrac{1}{2}\sqrt{n_2}\,\rho\tau b + c^2$$

$$= (n_1 + \sqrt{3n_1 n_2} + n_2)\rho^2 + c^2$$

and since $\rho^2 > 0, c^2 > 0$ we conclude that we must have

$$1 > n_1 + \sqrt{3n_1 n_2} + n_2. \tag{17.61}$$

Since n_1 and n_2 must take values from the set $(0, 1, 2, 3)$ the only option is $n_1 = n_2 = 0$, i.e. our additional simple root $\boldsymbol{\rho}$ is orthogonal to $\boldsymbol{\sigma}$ and $\boldsymbol{\tau}$ and our attempt at G_3 resulted in $G_2 \oplus A_1$.

f) In fact, Cartan showed that the only algebras possible are the classical ones—the infinite series A_n, B_n, C_n, D_n—and five exceptional algebras: G_2, F_4 and E_6, E_7 and E_8. We shall establish this result in the next chapter by making use of Dynkin diagrams. To finish this chapter we describe F_4 and the E_n series.

The simple roots of F_4 are

$$e_1-e_2-e_3-e_4, \quad e_2-e_3, \quad e_3-e_4, \quad 2e_4, \qquad (17.62)$$

all roots: 24 of the form $\quad \pm e_i \pm e_j, \quad 1 \leq i < j \leq 4,$

16 of the form $\quad \pm e_1 \pm e_2 \pm e_3 \pm e_4, \qquad (17.63)$

8 of the form $\quad \pm 2e_i,$

for a total of 48 non-zero roots. So the dimension of F_4 is 52. There are 24 short roots and 24 long roots, therefore

$$e_k \cdot e_j = \delta_{kj}/6. \qquad (17.64)$$

The fundamental weights are

$$2e_1, \quad e_1+e_2, \quad 2e_1+e_2+e_3, \quad 2(3e_1+e_2+e_3+e_4) \qquad (17.65)$$

and the dimension of the smallest representation is 26.

The E_n series is actually defined for $3 \leq n \leq 8$ but for $3 \leq n \leq 5$ the E_n are isomorphic to certain classical groups, as we will make clear in the next chapter.

The simple roots of E_n are

$$\sigma^{(1)} = -\tfrac{1}{2}(\sum_{j=2}^{n} e_j + \sqrt{9-n}\ e_1),$$

$$\sigma^{(2)} = e_2 + e_3, \qquad (17.66)$$

$$\sigma^{(j)} = -e_{j-1} + e_j, \quad 3 \leq j \leq n \leq 8,$$

all roots: $2(n-1)(n-2)$ of the form $\pm e_i \pm e_j, \quad 2 \leq i \neq j \leq n,$

2^{n-1} of the form $\quad \tfrac{1}{2}(\underbrace{\pm\sqrt{9-n}e_1 \pm e_2 \pm e_3 ... \pm e_n}_{even\ number\ of\ +\ signs}). \qquad (17.67)$

Further, E_7 has the additional two roots $\pm\sqrt{2}e_7$, and E_8 has the additional 28 roots $\pm e_1 \pm e_j, 1 \leq j \leq 7$.

It follows that the dimensions of the E_n and the normalizations of the e_j are as given in Table 17.2.

Table 17.2 Dimensions of E_n

n	dim of E_n	norm of e_j
3	11	
4	24	$10^{-1/2}$
5	45	$1/4$
6	78	$24^{-1/2}$
7	133	$1/6$
8	248	$60^{-1/2}$

We do not list the norm for $n=3$ because, as we shall see in the next chapter, $E_3 \cong A_1 \oplus A_2$ and so has roots of different length.

We saw in Chapter 10 that G_2 is a subgroup of $SO(7)$. We show this here by considering the respective root spaces. The root space of $SO(7)$ is three-dimensional with simple roots [see (17.36) and (17.37)] e_3, e_1-e_2, e_2-e_3, $e_j \cdot e_k = \delta_{jk}/10$. The long roots lie in a plane perpendicular to $e_1+e_2+e_3$. The projection of the short root e_3 onto that plane is $(2e_3-e_1-e_2)/3$. Thus, the simple roots in this plane are $\sigma=(2e_3-e_1-e_2)/3$ and $\tau=e_2-e_3$ since the third root e_1-e_2 is equal to $-(3\sigma+2\tau)$ and so is not simple. An easy calculation shows that $2\sigma \cdot \tau = -\sqrt{3}\sigma\tau$, that is these are the simple roots of G_2. Thus, the projection of the 3-dimensional root space of $SO(7)$ onto the 2-dimensional plane orthogonal to $e_1+e_2+e_3$ yields the root space of G_2.

18
Dynkin diagrams

It should be clear that all the information about an algebra is encoded in the simple roots. Cartan collected all that information into the so-called **Cartan matrix**, whose matrix elements are

$$A_{kl}=2\sigma^{(k)}\cdot\sigma^{(l)}/\sigma^{(l)}\cdot\sigma^{(l)}, \tag{18.1}$$

where the $\sigma^{(k)}$ are simple roots. The diagonal elements of the Cartan matrix equal 2, the off-diagonal elements, known as the **Cartan integers**, equal 0, -1, -2, or -3. It can be shown that because the root space is Euclidean the determinant of the Cartan matrix is positive and that in turn means that only certain choices of sets of simple roots are allowed, namely those corresponding to the algebras in the preceding chapter. We shall not prove this but instead consider certain diagrams invented by Dynkin when he was 19 years old and will prove that the only allowed diagrams are those corresponding to the algebras in the preceding chapter. The proof, again, depends on the root space being Euclidean.

Let $\boldsymbol{u}^i$ be a unit vector in the direction of the simple root $\sigma^{(i)}$:

$$\boldsymbol{u}^i=\sigma^{(i)}/\sigma^{(i)}, \tag{18.2}$$

so that our magic formula reads for $i\neq j$

$$\boldsymbol{u}^i\cdot\boldsymbol{u}^j=-\sqrt{\tfrac{p}{4}}, \quad p=0,1,2,3. \tag{18.3}$$

Now construct the **Dynkin diagram** as follows

1) Represent each unit vector $\boldsymbol{u}^i$ by a circle. ◯
2) Connect the circles corresponding to $\boldsymbol{u}^i$ and $\boldsymbol{u}^j$ by p lines with p given above; i.e. with *no* lines if $\boldsymbol{u}^i$ and $\boldsymbol{u}^j$ are at 90°, with *one* line if at 120°, with *two* lines if at 135°, with *three* lines if at 150°.
3) Indicate the relative length of the simple roots in some fashion.

Thus, for rank one we have a Dynkin diagram consisting of a single circle representing A_1.

For rank two we have

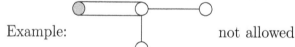

the relative length of roots being indicated by coloring the short root grey.

The algebra D_2 is the direct sum of two A_1 algebras. For the corresponding groups $SO(4) \cong SU(2) \otimes SU(2)$, that is $SO(4)$ is the direct product of two $SU(2)$ groups. In general, then, if a Dynkin diagram contains disconnected pieces we are dealing with a semisimple algebra corresponding to a direct sum of simple algebras. This is because, by construction, all the roots in one of the disconnected parts are orthogonal to all the roots in the other disconnected part (every generator in one part commutes with every generator in the other part). In what follows we consider mainly diagrams that are *connected* corresponding to *simple* algebras.

We start by stating three features that a Dynkin diagram must satisfy:

A) There are no loops. Example: ▽ not allowed.

B) No more than three lines can emanate from a circle.

Example: ⌐ not allowed

C) A set of circles u^i connected to one another by a single line can be shrunk to a single circle u. The result is an allowed diagram if the original is allowed (and the original is not allowed if the result is not allowed).

We shall prove these properties shortly but first let us look at the consequences. By property B we can have at most three lines attached to a circle, which clearly can only be done in the following three ways:

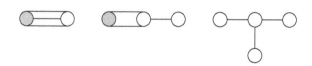

Dynkin diagrams

The first of these is G_2 and it *cannot* be attached to another circle because then there would be four lines emanating from a circle. Thus, three lines connecting two circles occur only for G_2 and we can forget about it from now on (this immediately explains why G_2 did not generalize to higher rank, in contrast to A_2 or B_2).

Now what about the other two diagrams above? Can we use them twice in a larger diagram? The answer is no, because they would have to be connected by a chain of circles connected by single lines, which could be shrunk away and again we would have a circle with four lines emanating from it.

Thus, we conclude that the only allowed Dynkin diagrams, in addition to the one for G_2, are

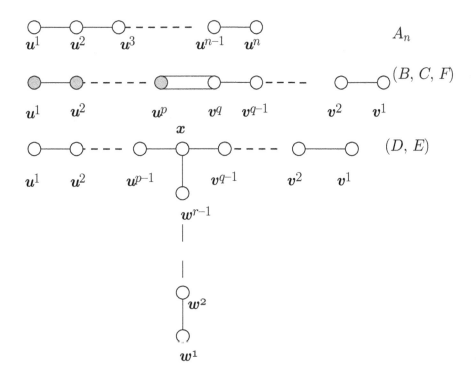

For the (B,C,F) diagram we shall now prove that the only allowed values for p,q are: $q=1, p$ arbitrary; $p=1, q$ arbitrary; and $p=q=2$. Define

$$u = \sum_{k=1}^{p} k u^k, \quad v = \sum_{l=1}^{q} l v^l \qquad (18.4)$$

so that

$$u \cdot u = \sum_{k=1}^{p} k^2 u^k \cdot u^k + 2\sum_{k=1}^{p-1} k(k+1) u^k \cdot u^{k+1}$$

$$= \sum_{k=1}^{p} k^2 - \sum_{k=1}^{p-1} k(k+1) = p(p+1)/2 \qquad (18.5)$$

and

$$u \cdot v = p u^p \cdot q v^q = -pq/\sqrt{2}. \qquad (18.6)$$

We now use Schwartz's inequality (remember: Euclidean geometry)

$$(u \cdot v)^2 < (u \cdot u)(v \cdot v),$$

i.e.

$$p^2 q^2 / 2 < p(p+1)q(q+1)/4. \qquad (18.7)$$

Since by construction p and q are non-zero we can rewrite above as

$$2 < (1+\tfrac{1}{p})(1+\tfrac{1}{q}). \qquad (18.8)$$

Suppose $q=1$, then we must have $1 \leq 1+\tfrac{1}{p}$, which is true for arbitrary $p \geq 1$. The same result holds with p and q interchanged. Next suppose that $q>1$ and $p>1$, then (18.8) is valid only for $p=q=2$. The corresponding Dynkin diagrams are

$B_{p+1} = so(2(p+1)+1)$

$C_{p+1} = sp(2(p+1))$

F_4

Lastly, consider the (D,E) diagram. Here, we will show that allowed values of p, q, r, in that diagram are (assuming without loss of generality $p \geq q \geq r$) as shown in Table 18.1.

Table 18.1 Allowed values of p, q and r

p	q	r	algebra
arbitrary	2	2	D series
3	3	2	E_6
4	3	2	E_7
5	3	2	E_8

Define

$$\boldsymbol{u}=\sum_{k=1}^{p-1}k\boldsymbol{u}^k, \quad \boldsymbol{v}=\sum_{l=1}^{q-1}l\boldsymbol{v}^l, \quad \boldsymbol{w}=\sum_{s=1}^{r-1}s\boldsymbol{w}^s, \tag{18.9}$$

so that

$$\boldsymbol{u}\cdot\boldsymbol{u}=p(p-1)/2, \quad \boldsymbol{v}\cdot\boldsymbol{v}=q(q-1)/2, \quad \boldsymbol{w}\cdot\boldsymbol{w}=r(r-1)/2 \tag{18.10}$$

and by construction the smallest non-trivial value of p, q, r, is 2.
The unit vector $\boldsymbol{x}$, from which emanate three lines, satisfies

$$\boldsymbol{x}\cdot\boldsymbol{x}=1,$$
$$\boldsymbol{x}\cdot\boldsymbol{u}=(p-1)\;\boldsymbol{x}\cdot\boldsymbol{u}^{p-1}=-(p-1)/2, \tag{18.11}$$
$$\boldsymbol{x}\cdot\boldsymbol{v}=(q-1)\boldsymbol{x}\cdot\boldsymbol{v}^{q-1}=-(q-1)/2,$$
$$\boldsymbol{x}\cdot\boldsymbol{w}=(r-1)\boldsymbol{x}\cdot\boldsymbol{w}^{r-1}=-(r-1)/2$$

and

$$\cos^2(\boldsymbol{x},\boldsymbol{u})=(\boldsymbol{x}\cdot\boldsymbol{u})^2/(\boldsymbol{x}\cdot\boldsymbol{x})(\boldsymbol{u}\cdot\boldsymbol{u})=\frac{(p-1)^2/4}{p(p-1)/2}=\frac{1}{2}\left(1-\frac{1}{p}\right). \tag{18.12}$$

Since $\boldsymbol{x}$, $\boldsymbol{u}$, $\boldsymbol{v}$, $\boldsymbol{w}$ are linearly independent and $\boldsymbol{x}$ is a unit vector the sum of squares of direction cosines of $\boldsymbol{x}$ along the directions $\boldsymbol{u}$, $\boldsymbol{v}$, $\boldsymbol{w}$ is less than 1:

$$\tfrac{1}{2}(1-\tfrac{1}{p}+1-\tfrac{1}{q}+1-\tfrac{1}{r})<1 \quad \text{or} \quad \tfrac{1}{p}+\tfrac{1}{q}+\tfrac{1}{r}>1. \tag{18.13}$$

This Diophantine inequality has the following solutions: clearly if $p=q=r=3$ the inequality is violated and larger values for $p=q=r$ only make matters worse. Since $p\geq q\geq r\geq 2$ it follows that for any solution to exist we must have $r=2$ and (18.13) becomes

$$\tfrac{1}{p}+\tfrac{1}{q}>\tfrac{1}{2} \tag{18.14}$$

with solutions given by Table 18.1. The corresponding Dynkin diagrams are

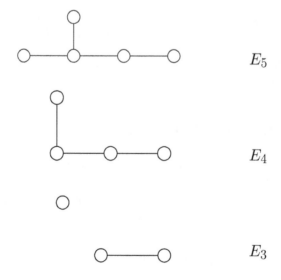

We have renamed the roots in E_8 to correspond to the naming of the simple roots of E_8 in (17.66). Then the diagram for E_7 results by omitting the leftmost root in E_8, the diagram for E_6 results by omitting the leftmost root in E_7. Continuing in this vein we get

We have hinted before that the $E_{5,4,3}$ are not new. Indeed, as we look at these Dynkin diagrams and as is confirmed by the dimensions as given in Table 17.2 we have

$$E_3 = A_1 \oplus A_2, \quad E_4 = A_4, \quad E_5 = D_5. \tag{18.15}$$

It is noteworthy that this sequence mirrors almost exactly the algebras used in particle physics as the Standard Model $U(1) \otimes SU(2) \otimes SU(3)$, its grand unification $SU(5)$, and later $Spin(10)$. It suggests that particle physics might ultimately be described by E_8, an algebra that is somewhat unique by being the largest exceptional algebra.

The ADE algebras are collectively referred to as **simply laced**, meaning that all the roots are of the same length, all the lines in the Dynkin diagram are single. The A-D-E classification arises in a remarkable variety of situations:

1. Weyl groups with roots of equal length;
2. finite groups generated by reflections;
3. the platonic solids;
4. representations of quivers;
5. singularities of algebraic hypersurfaces with a definite intersection form;
6. critical points of functions having no moduli.

This completes the classification of all simple algebras using the Dynkin method except that we need to prove the three features of Dynkin diagrams stated earlier.

A) There are no loops.

Suppose that the Dynkin diagram consists of a polygon with k vertices. Label the circle at each vertex by $\boldsymbol{u}^i$, a unit vector in the direction of a simple root. Construct $\boldsymbol{u} = \boldsymbol{u}^1 + \boldsymbol{u}^2 + \boldsymbol{u}^3 + \ldots + \boldsymbol{u}^k$. Since the simple roots are linearly independent, $\boldsymbol{u} \neq 0$. Therefore,

$$0 < \boldsymbol{u} \cdot \boldsymbol{u} = \left(\sum_{i=1}^{k} \boldsymbol{u}^i\right) \cdot \left(\sum_{i=1}^{k} \boldsymbol{u}^i\right) = \sum_{i=1}^{k} \boldsymbol{u}^i \cdot \boldsymbol{u}^i + 2\sum_{i=1}^{k} \boldsymbol{u}^i \cdot \boldsymbol{u}^{i+1} \tag{18.16}$$

(where $\boldsymbol{u}^{k+1} = \boldsymbol{u}^1$). If the neighboring roots are connected by a single line $2\boldsymbol{u}^i \cdot \boldsymbol{u}^{i+1} = -1$ and the right-hand side of (18.16) is zero, if some neighbors are connected by more than one line the right-hand side of

(18.16) is negative—either way the inequality is violated and this proves that loops are not allowed.

B) No more than three lines can emanate from a circle.

To prove this feature we need the following result first. Suppose that $v^1, v^2, \ldots$ form an *orthonormal* system

$$v^i \cdot v^j = \delta^{ij}. \tag{18.17}$$

Let u be a unit vector. Then $u \cdot v^i$ is the direction cosine of u along v^i and

$$\sum_i (u \cdot v^i)^2 = \sum_i \cos^2(u, v^i) \leq 1 \tag{18.18}$$

and for the equality to hold must have $u = \sum_i v^i$, i.e. u and the v^i must be linearly dependent.

Now, consider a Dynkin diagram and denote by u the root corresponding to a particular circle, by v^i all the roots that are connected to u. All the v^i are unit vectors by construction and they are never connected to each other because loops are not allowed—hence the v^i form an orthonormal set and we have

$$\sum_i (u \cdot v^i)^2 = \sum_i p_i/4 \leq 1, \tag{18.19}$$

where the p_i indicate the number of lines between u and v^i, hence we get

$$\sum_i p_i < 4 \tag{18.20}$$

for u and v^i linearly independent.

C) A set of circles u^i connected to one another by a single line can be shrunk to a single circle u. The result is an allowed diagram if the original is allowed (and the original is not allowed if the result is not allowed).

As an example, consider the diagram

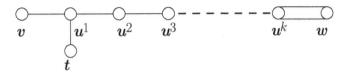

and define

$$u = \sum_{i=1}^{k} u^i. \tag{18.21}$$

Then,

$$u \cdot u = \sum_{i=1}^{k} u^i \cdot u^i + 2\sum_{i=1}^{k-1} u^i \cdot u^{i+1} = k - (k-1) = 1, \tag{18.22}$$

$$v \cdot u = v \cdot \sum_{i=1}^{k} u^i = v \cdot u^1, \; t \cdot u = t \cdot \sum_{i=1}^{k} u^i = t \cdot u^1, \; w \cdot u = w \cdot \sum_{i=1}^{k} u^i = w \cdot u^k, \tag{18.23}$$

so we see that the original diagram is not allowed because the shrunk diagram below involves the same data (for unshrunk roots) and is not allowed

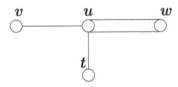

Since all the information about the algebra is encoded in the Dynkin diagram two algebras with the same diagram must be isomorphic. In addition to $A_1 = B_1 = C_1$, $B_2 = C_2$, $D_2 = A_1 \oplus A_1$, as well as the isomorphisms involving $E_{3,4,5}$, we have $D_3 = A_3$.

In previous chapters we have discussed automorphisms of the $so(n)$ and $su(n)$ algebras. If a Dynkin diagram can be mapped into itself in a way that preserves the inner products of the simple roots we will obtain an automorphism of the algebra. This happens if the diagram has some symmetries. It is fairly obvious that the non-simply laced diagrams (i.e. B_n, C_n, F_4 and G_2) have no symmetries. The A_n diagram is symmetric upon being flipped over, i.e. under the exchange of the ith and $(n+1-i)$th circle. This corresponds to complex conjugation, as was discussed in the chapter on unitary groups. We noted there that this was not applicable to $SU(2)$ because of its ambivalence—of course the Dynkin diagram for $SU(2) = A_1$ consists of a single circle and the symmetry under flipping is empty of content in this case. The E_6 diagram exhibits a similar symmetry under flipping. The D_n diagrams are symmetric under the exchange of the two circles in the fork. Those two circles describe semispinors and this automorphism corresponds to reflection. Finally, the D_4 diagram exhibits a symmetry of order three (all the previous ones are of order two) involving the exchange of any of its three legs that describe the two semispinors and the adjoint, all three representations being 8-dimensional and real, this automorphism is called *triality*.

These symmetries of the Dynkin diagrams lead to a phenomenon called **folding**. We start with D_4 and draw its Dynkin diagram in the suggestive way

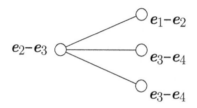

The three roots on the right are permuted under the automorphism of this diagram. The folding consists of summing them into a new root

$$(e_1-e_2)+(e_3-e_4)+(e_3+e_4)=e_1-e_2+2e_3$$

giving rise to the Dynkin diagram

$$e_2\text{-}e_3 \; \Longleftarrow \; e_1\text{-}e_2+2e_3$$

and although the e_1, e_2 and e_3 span a 3-dimensional space the two simple roots of G_2 live in the 2-dimensional space orthogonal to $e_1+e_2+e_3$.

The A_{2n-1} diagram is symmetric under the exchange of roots $\sigma^{(i)}$ and $\sigma^{(2n-i)}$, $1 \leq i \leq n-1$, with the root $\sigma^{(n)}$ left alone. The folding consists of replacing the roots that go into each other by their sum and the resultant diagram is that of B_n—we show this below by folding A_5 into B_3:

The D_{n+1} diagram is symmetric under the exchange of the two roots in the fork—replacing those two roots by their sum folds D_{n+1} into C_n as we demonstrate below by folding D_5 into C_4:

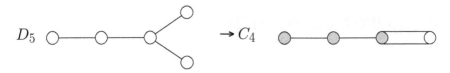

Finally, the automorphism of E_6 makes it possible to fold E_6 into F_4:

Thus, all the non-simply laced algebras can be obtained from the simply laced ones by folding.

19
The Lorentz group

Our interest in groups has to do among other things with the fact that there are certain *symmetries* in space-time and we can associate groups with them. What we have in mind here is the following: consider a non-relativistic force problem with the force being a so-called central force—in such a problem there occurs a constant of the motion known as *angular momentum*, and the solution of the problem is considerably simplified once that is recognized.

How does this relate to symmetry? If $\boldsymbol{L}$ denotes the angular momentum operator then the fact that it is a constant of the motion means that it is invariant under translations in time. That means that it commutes with the generator of translations in time, which is the Hamiltonian H:

$$[H, \boldsymbol{L}]=0. \tag{19.1}$$

However, (19.1) can equally well be read as the statement that H is left invariant under transformations generated by $\boldsymbol{L}$ that, of course, are rotations and, as everybody knows H for a central force problem is *spherically symmetric*.

Thus, we arrive at our first and most elementary symmetry group of space-time, namely $SO(3)$: rotations in the three-dimensional space of everyday life. This is a group that we have discussed before and it is well understood. Now we may ask this question: is there perhaps *more* symmetry to the equations of physics than just the rotational symmetry mentioned above? If so, we must look for a larger group that contains $SO(3)$ as a subgroup.

The Lorentz group

We contemplate the generalization suggested by *special relativity*, namely what is called the **Lorentz group**. The theory of special relativity can be thought of as the statement that the laws of physics are *invariant* under rotations in a *four-dimensional space-time*. Using x_μ, $1\le\mu\le 4$, to denote the four coordinates $x_1=x$, $x_2=y$, $x_3=z$, $x_4=ict$ (where c is the velocity of light, which we shall set equal to 1 from now on) we observe that the invariance of the quadratic form

$$x_\mu x_\mu \qquad (19.2)$$

is guaranteed for transformations generated by the six operators $M_{\mu\nu} = -M_{\nu\mu}$ corresponding to the orthogonal group in four dimensions. By convention when discussing special relativity lower case Greek letters denote indices that range from 1 to 4, lower case Latin letters denote indices that range from 1 to 3. Using an imaginary fourth coordinate for time to make Lorentz transformations look like rotations is due to Minkowski. We should note, however, that when we express the quadratic form in terms of the real variables $\boldsymbol{x}$ and $x_0=t$ rather than $x_4=ix_0$ we have

$$x_\mu x_\mu = \boldsymbol{x}\cdot\boldsymbol{x} - x_0^2, \qquad (19.3)$$

which means that the Lorentz group is $O(3,1)$ [see (12.8)].

In particular, it follows that $O(3,1)$ transformations consist of four separate components: the transformations are *proper* if the determinant is $+1$, *improper* if the determinant is -1, they are *orthochronous* if they preserve the direction of time, *non-orthochronous* if they reverse the direction of time. It is the proper orthochronous component that is connected to the identity and so forms a group, which we denote by $SO_0(3,1)$.

The most crucial consequence of the indefinite metric is that the Lorentz group is *non-compact*. To see this, consider the six parameters that are needed to parameterize the Lorentz transformations. The generators M_{ij} generate of course the subgroup $SO(3)$—a fine compact group that can be parameterized by the three Euler angles, resulting in a finite volume in parameter space.

The three remaining generators $M_{j4} = -M_{4j} = -iM_{0j}$ generate rotations in the $j4$-plane. These are pure Lorentz transformations, the so-called **boosts**, which are transformations from one coordinate system to another, the two systems moving relative to each other with constant velocity in the jth direction, that is the remaining three parameters are the three components of the relative velocity. There is a clever way of expressing this as a rotation but because the $j4$-plane has a real and an imaginary axis the angle of rotation is imaginary, trigonometric functions describing the transformations become hyperbolic and volume in parameter space becomes infinite.

Now all this has profound consequences. Consider the six generators of $SO(3,1)$. If we write a group element as

$$g=\exp(ir_{\mu\nu}M_{\mu\nu})=\exp(ir_{jk}M_{jk}+ir_{4j}M_{4j}) \qquad (19.4)$$

the $r_{jk}=-r_{kj}$ are three *real* parameters but the r_{4j} are, because of the Minkowski trick, three pure *imaginary* parameters. Therefore, the M_{jk} are hermitian and the M_{4j} are anti-hermitian in a *unitary* representation.

We remind the reader that we have shown previously that for non-compact groups unitary representations are infinite-dimensional, while finite-dimensional representations are non-unitary. So let's look at representations of the Lorentz group. Starting from the realization of the generators as

$$M_{\mu\nu}=-ix_{[\mu}\partial_{\nu]} \qquad (19.5)$$

we get the commutation relations

$$[M_{\mu\nu},M_{\alpha\beta}]=i(\delta_{\alpha[\mu}M_{\nu]\beta}-\delta_{\beta[\mu}M_{\nu]\alpha}). \qquad (19.6)$$

We take (19.6) as the definition of the Lorentz agebra whether the $M_{\mu\nu}$ are realized as in (19.5) or not.

If we define the components of the angular momentum $\boldsymbol{J}$ and the boost $\boldsymbol{N}$ by

$$J_j=\tfrac{1}{2}\varepsilon_{jkl}M_{kl}, \quad N_j=iM_{4j} \qquad (19.7)$$

then their commutation relations are

$$[J_j,J_k]=i\varepsilon_{jkl}J_l, \qquad (19.8)$$

$$[J_j,N_k]=i\varepsilon_{jkl}N_l, \qquad (19.9)$$

$$[N_j,N_k]=-i\varepsilon_{jkl}J_l, \qquad (19.10)$$

and in a unitary representation J_k and N_k are hermitian.

We see that (19.8) identifies the components of $\boldsymbol{J}$ as the generators of the $SO(3)$ subgroup, and (19.9) identifies $\boldsymbol{N}$ as a vector (tensor of rank one) with respect to the rotations generated by $\boldsymbol{J}$. Equation (19.10) states that the components of $\boldsymbol{N}$ do *not* generate a subgroup and ultimately results in the Thomas precession phenomenon. We can also immediately identify, recalling our discussion of orthogonal groups, the two Casimir operators of $SO(3,1)$ (here $\varepsilon_{\alpha\beta\mu\nu}$ is the totally antisymmetric invariant tensor with $\varepsilon_{1234}=+1$)

$$F_1 = \tfrac{1}{2} M_{\alpha\beta} M_{\alpha\beta} = \boldsymbol{J} \cdot \boldsymbol{J} - \boldsymbol{N} \cdot \boldsymbol{N}, \qquad (19.11)$$

$$F_2 = \tfrac{1}{4} \varepsilon_{\alpha\beta\mu\nu} M_{\alpha\beta} M_{\mu\nu} = 2\mathrm{i} \boldsymbol{J} \cdot \boldsymbol{N}. \qquad (19.12)$$

(Manifestly these are invariant under the $SO(3)$ subgroup as they are given by scalar products of three-vectors.)

We can form a canonical basis in terms of the states $|jm\rangle$ discussed in Chapter 4:

$$\boldsymbol{J} \cdot \boldsymbol{J} |jm\rangle = j(j+1)|jm\rangle, \quad J_3|jm\rangle = m|jm\rangle,$$

$$-j \leq m \leq j, \quad 2j = 0,1,2,.... \qquad (19.13)$$

We have included half-odd-integer values for j because we are actually considering representations not of $SO_0(3,1)$ but of its universal covering group, which is $SL(2,\mathbb{C})$. In particular, this means that instead of the $SO(3)$ subgroup of $SO(3,1)$ we are dealing with its universal cover, namely $SU(2)$.

While we leave the demonstration that $SL(2,\mathbb{C})$ is the universal cover of $SO_0(3,1)$ for later we explain now the need for it. In quantum mechanics one is dealing with so-called *projective* representations that are representations up to a phase. This means that we are free to choose phases to satisfy some other requirements such as continuity. It turns out that for semisimple Lie groups we are free to make such choices *locally* and these choices will be valid *globally* provided the group is simply connected. This is achieved (see end of Chapter 3) by going to the universal covering group.

Now, it is clear that we are contemplating here the chain $SL(2,\mathbb{C}) \downarrow SU(2)$ and the question is: which representations of $SU(2)$ are contained in a representation of $SL(2,\mathbb{C})$. The answer, of course, involves the quadratic Casimirs $F_{1,2}$ whose evaluation requires the knowledge of the matrix elements of $\boldsymbol{J}$ and $\boldsymbol{N}$ in the above basis.

We know that for $\boldsymbol{J}$ the only non-vanishing matrix elements are

$$\langle jm|J_3|jm\rangle = m, \quad \langle jm|J_\pm|jm\mp 1\rangle = \sqrt{(j\pm m)(j\mp m+1)}. \qquad (19.14)$$

These relations exhaust the content of the commutation relations (9.8). The commutation relations (9.9) are the statement that $\boldsymbol{N}$ is a tensor of rank one so that we have from the Wigner–Eckart theorem (see Chapter 4) that

$$\langle j'm'|N_\lambda|jm\rangle = \langle j'm'|1\lambda,jm\rangle\langle j'\|N\|j\rangle, \qquad (19.15)$$

where we denote the reduced matrix elements by $\langle j'\|N\|j\rangle$. The Clebsch–Gordan coefficient vanishes except when $m'=m+\lambda$ and $j'=j-1, j$, or $j+1$. Thus, in contrast to $\boldsymbol{J}$, $\boldsymbol{N}$ connects different irreducible representations of the $SU(2)$ generated by $\boldsymbol{J}$. In particular, starting from some j we can use $\boldsymbol{N}$ to obtain $j-1$, and then $j-2$, and so on. This procedure ends when we reach $j=0$ or $j=1/2$. It can also end at some value to be called j_0 if

$$\langle j_0-1\|N\|j_0\rangle = 0. \qquad (19.16)$$

Using $\boldsymbol{N}$ we can also *raise* the values of j, but while there is a minimum value of j there is no maximum value since a unitary representation is infinite-dimensional. The possible values of j_0 are, of course, $0, 1/2, 1, 3/2, \ldots$. Thus, an irreducible representation of $SL(2,\mathbb{C})$ corresponding to j_0 equal to some integer (half-odd-integer) contains all integer (half-odd-integer) values of j larger or equal to j_0.

So far, we have only used the fact that $\boldsymbol{N}$ is a tensor of rank one. It is in fact a particular tensor of rank one, namely it satisfies in addition the commutation relations (19.10), which can be used to determine the reduced matrix elements. We shall skip the details of that calculation and state the results: the Casimir operators can be written as

$$F_1 = (j_0^2 - 1 + \rho^2), \quad F_2 = j_0 \rho. \qquad (19.17)$$

Recalling that $F_1 = (\boldsymbol{J}^2 - \boldsymbol{N}^2)/2$, $F_2 = i\boldsymbol{J}\cdot\boldsymbol{N}$, it follows that in a unitary representation F_1 is real and F_2 is pure imaginary or zero. Thus, we have the so-called **principal series** for which

$$2j_0 = 0, 1, 2, \ldots, \quad -\infty \leq i\rho \leq \infty, \qquad (19.18)$$

and the **complementary series** for which

$$j_0 = 0, \quad -1 \leq \rho \leq 1. \qquad (19.19)$$

Here, F_1 imposes no restrictions on the magnitude of ρ in the principal series, while in the complementary series that magnitude must not exceed one to ensure that F_1 is negative since

$$F_1=\langle j_0 j_0|(\boldsymbol{J}^2-\boldsymbol{N}^2)|j_0 j_0\rangle/2\xrightarrow[j_0=0]{}-\langle 00|\boldsymbol{N}^2|00\rangle/2. \quad (19.20)$$

Why do we insist on unitary representations? Well, the operators J_i and N_i are so-called *observables*, meaning that any one (or several, if they commute) can be (simultaneously) measured in an experiment and the result of a measurement is, on the one hand, an eigenvalue of the operator in question and, on the other hand, a real number. Thus, we require for observables operators whose eigenvalues are real. Hermitian operators have real eigenvalues and J_i and N_i are hermitian in a unitary representation.

In contrast to J_i and N_i the wave function satisfying a relativistic wave equation is not an observable and so can be described by finite-dimensional (non-unitary) representations of the $SL(2,\mathbb{C})$ group. To conclude the discussion of the Lorentz group we describe these representations. If we form out of $\boldsymbol{J}$ and $\boldsymbol{N}$ the combinations

$$\boldsymbol{A}=(\boldsymbol{J}+\mathrm{i}\boldsymbol{N})/2, \quad \boldsymbol{B}=(\boldsymbol{J}-\mathrm{i}\boldsymbol{N})/2 \quad (19.21)$$

we find that the commutation relations (19.8) – (19.10) become

$$[A_i, A_j]=\mathrm{i}\varepsilon_{ijk}A_k,$$
$$[B_i, B_j]=\mathrm{i}\varepsilon_{ijk}B_k, \quad (19.22)$$
$$[A_i, B_j]=0.$$

Moreover, we have

$$\boldsymbol{J}=\boldsymbol{A}+\boldsymbol{B}, \quad (19.23)$$
$$\boldsymbol{A}^2=(\boldsymbol{J}^2-\boldsymbol{N}^2+2\mathrm{i}\boldsymbol{J}\cdot\boldsymbol{N})/4, \quad \boldsymbol{B}^2=(\boldsymbol{J}^2-\boldsymbol{N}^2-2\mathrm{i}\boldsymbol{J}\cdot\boldsymbol{N})/4, \quad (19.24)$$

and therefore

$$F_1=\boldsymbol{A}^2+\boldsymbol{B}^2, \quad F_2=\boldsymbol{A}^2-\boldsymbol{B}^2. \quad (19.25)$$

Now, let $\boldsymbol{J}$ be represented by a hermitian operator, $\boldsymbol{N}$ by an anti-hermitian operator, hence of course we have a *non-unitary* representation of the Lorentz group. It follows that $\boldsymbol{A}$ and $\boldsymbol{B}$ are hermitian, we are exploiting the isomorphism $Spin(4)\cong SU(2)\otimes SU(2)$ and the commutation relations (19.22) yield in the standard manner the results

$$\boldsymbol{A}^2 = j_a(j_a+1), \quad \boldsymbol{B}^2 = j_b(j_b+1),$$
$$2j_a = 0,1,2,3,..., \quad 2j_b = 0,1,2,3,.... \tag{19.26}$$

These $(2j_a+1)(2j_b+1)$-dimensional representations are usually denoted by (j_a, j_b). They are manifestly finite-dimensional and since $\boldsymbol{J} = \boldsymbol{A} + \boldsymbol{B}$ the j-content of these representations is given by

$$|j_a - j_b| \leq j \leq j_a + j_b. \tag{19.27}$$

It is useful here to introduce the **parity** operation defined by the following action on the position four-vector:

$$x_i \to -x_i, \quad x_4 \to +x_4. \tag{19.28}$$

Thus, under the parity transformation the generators of the Lorentz group behave as follows: $M_{ij} \to M_{ij}, M_{i4} \to -M_{i4}$, i.e. $\boldsymbol{J} \to \boldsymbol{J}$, $\boldsymbol{N} \to -\boldsymbol{N}$ (thus $\boldsymbol{N}$ is a polar or true vector, $\boldsymbol{J}$ is an axial or pseudovector) and so $\boldsymbol{A} \to \boldsymbol{B}$, $\boldsymbol{B} \to \boldsymbol{A}$ and finally $(j_a, j_b) \to (j_b, j_a)$. Consequently, these finite-dimensional representations can be irreducible representations of the Lorentz group extended by parity provided they are either of the form

$$(j_a, j_b), \quad j_a = j_b, \tag{19.29}$$

or of the form

$$(j_a, j_b) \oplus (j_b, j_a), \quad j_a \neq j_b. \tag{19.30}$$

The first kind is $(2j_a+1)^2$-dimensional, the second kind is $2(2j_a+1)(2j_b+1)$-dimensional.

The $(0,0)$ case: This one-dimensional representation is used in the Klein–Gordon equation and obviously describes spin 0.

The $(1/2,0) \oplus (0,1/2)$ case: This four-dimensional representation is used in the Dirac equation; it describes spin 1/2 twice (a particle and its antiparticle).

The $(1/2,0)$ or $(0,1/2)$ separately case: These describe spin 1/2 particles individually and are used in the Weyl equation that violates invariance under parity, as well as under particle–antiparticle conjugation.

The $(1/2,1/2)$ case: This four-dimensional representation is used in the Proca equation; it describes a four-vector A_ν that contains spin 0 and spin 1. The $j=0$ part is usually eliminated by requiring $\partial_\nu A_\nu = 0$.

The *(1,0)⊕(0,1) case:* This six-dimensional representation is used in the Maxwell equation; it describes $F_{\mu\nu}$, a second rank tensor in four dimensions containing spin 1 twice—the electric and magnetic fields.

To finish this chapter we prove that $SL(2,\mathbb{C})$ is the universal covering group of $SO_0(3,1)$, the proper orthochronous Lorentz group.

The group $SL(2,\mathbb{C})$

Consider the four matrices σ_μ:

$$\sigma_1=\begin{pmatrix}0 & 1\\ 1 & 0\end{pmatrix}, \quad \sigma_2=\begin{pmatrix}0 & -i\\ i & 0\end{pmatrix}, \quad \sigma_3=\begin{pmatrix}1 & 0\\ 0 & -1\end{pmatrix}, \quad \sigma_4=\begin{pmatrix}i & 0\\ 0 & i\end{pmatrix}, \qquad (19.31)$$

$$\mathrm{tr}\,\sigma_\mu^\dagger \sigma_\nu = 2\delta_{\mu\nu}. \qquad (19.32)$$

An arbitrary 2×2 matrix can be expanded in terms of the σ_μ since they span the space of 2×2 matrices. In particular, with x_μ denoting the distance between two events in space-time consider the matrix X:

$$X=\sigma_\mu x_\mu=\begin{pmatrix}ix_4+x_3 & x_1-ix_2\\ x_1+ix_2 & ix_4-x_3\end{pmatrix},$$

$$x_\mu=\tfrac{1}{2}\mathrm{tr}\,\sigma_\mu^\dagger X, \quad \mathrm{tr}\,X=2ix_4, \quad \det X=-x_\mu x_\mu. \qquad (19.33)$$

Note that X is hermitian because $\boldsymbol{x}$ is real and x_4 is pure imaginary.

We now introduce the $SL(2,\mathbb{C})$ group. Let $Q\in SL(2,\mathbb{C})$ and define X' by

$$X'=QXQ^\dagger. \qquad (19.34)$$

It follows that X' is hermitian so can be expressed in the same form as X with x_μ replaced by x'_μ. Thus, we have

$$x'_\mu=\tfrac{1}{2}\mathrm{tr}\,\sigma_\mu^\dagger X'=\tfrac{1}{2}\mathrm{tr}\,\sigma_\mu^\dagger QXQ^\dagger=\tfrac{1}{2}\mathrm{tr}\,\sigma_\mu^\dagger Q\sigma_\rho x_\rho Q^\dagger=\Lambda(Q)_{\mu\rho}x_\rho, \qquad (19.35)$$

where

$$\Lambda(Q)_{\mu\rho}=\tfrac{1}{2}\mathrm{tr}\,\sigma_\mu^\dagger Q\sigma_\rho Q^\dagger. \qquad (19.36)$$

Since Q is unimodular $\det X'=\det X$, so $x'_\mu x'_\mu=x_\mu x_\mu$ and $\Lambda(Q)_{\mu\rho}\in O(3,1)$. In fact, $\Lambda(Q)_{\mu\rho}\in SO(3,1)$ but the proof is tedious and we skip it. From (19.36) we get

$$\Lambda(Q)_{44}=\tfrac{1}{2}\text{tr }\sigma_4^\dagger Q\sigma_4 Q^\dagger=\tfrac{1}{2}\text{tr}QQ^\dagger>0 \tag{19.37}$$

so $\Lambda(Q)_{\mu\rho}$ is orthochronous, i.e. $\Lambda(Q)_{\mu\rho}\in SO_0(3,1)$. The correspondence between Q and $\Lambda(Q)_{\mu\rho}$ means that $SL(2,\mathbb{C})$ and $SO_0(3,1)$ are homomorphic. Since cQ and Q, where c is some complex number, give rise to the same $\Lambda(Q)_{\mu\rho}$ provided $cc^*=1$, and since cQ and Q are both unimodular we must actually have $c=\pm 1$ and we see that the homomorphism is 2:1 as $\pm Q$ correspond to the same $\Lambda(Q)_{\mu\rho}$. Thus, we have the isomorphism

$$SL(2,\mathbb{C})/\mathbb{Z}_2\cong SO_0(3,1). \tag{19.38}$$

Note that $SU(2)$ is a subgroup of $SL(2,\mathbb{C})$ and for $Q\in SU(2)$ (19.34) becomes a similarity transformation that preserves traces so that $x_4'=x_4$ and the corresponding Lorentz transformation $\Lambda(Q)_{\mu\rho}\in SO(3)$. Thus, we have as a special case of (19.38) the familiar

$$SU(2)/\mathbb{Z}_2\cong SO(3). \tag{19.39}$$

To complete the proof that $SL(2,\mathbb{C})$ is the universal cover of $SO_0(3,1)$ we must show that $SL(2,\mathbb{C})$ is simply connected.

Let $Q\in SL(2,\mathbb{C})$ and consider the matrix $Q^\dagger Q$. It is clearly unimodular and hermitian, it is also positive (meaning that all its eigenvalues are positive). To see that write the eigenvalue equation

$$Q^\dagger QD=dD \tag{19.40}$$

where d is one of the eigenvalues of $Q^\dagger Q$ and D is the corresponding eigenvector (a one-column matrix). It follows that

$$d\,D^\dagger D=D^\dagger Q^\dagger QD=(QD)^\dagger QD, \tag{19.41}$$

which shows that $d>0$ since $D^\dagger D>0$ and $(QD)^\dagger QD>0$. Since $Q^\dagger Q$ is positive we can take its square root and define a unimodular hermitian positive 2×2 matrix P by

$$P=(Q^\dagger Q)^{1/2}. \tag{19.42}$$

Lastly, we define a matrix U by

$$U=QP^{-1}. \tag{19.43}$$

It follows that

$$U^\dagger U = P^{-1} Q^\dagger Q P^{-1} = P^{-1} P^2 P^{-1} = 1, \quad \det U = +1, \tag{19.44}$$

that is $U \in SU(2)$.

Thus, we have the so-called **polar decomposition** of an arbitrary matrix $Q \in SL(2,\mathbb{C})$ as the product of a unimodular unitary matrix U and a unimodular hermitian positive matrix P

$$Q = UP, \tag{19.45}$$

in complete analogy to the decomposition of a complex number

$$z = \exp i\varphi (z^* z)^{1/2}. \tag{19.46}$$

Since P is unimodular hermitian and positive it can be written as

$$P = \exp A, \tag{19.47}$$

where A is hermitian and traceless, hence can be parameterized in terms of three completely unrestricted real numbers, that is the parameter space of P is $\mathbb{R}^3$. Since $U \in SU(2)$, its parameter space is $\mathbb{S}^3$ and we conclude that the parameter space of $SL(2,\mathbb{C})$ is the union of $\mathbb{S}^3$ and $\mathbb{R}^3$ and therefore $SL(2,\mathbb{C})$ is simply connected, since $\mathbb{S}^n$ for $n > 1$ and $\mathbb{R}^n$ for all n are all simply connected.

This completes the demonstration that $SL(2,\mathbb{C})$ is the universal covering group of $SO_0(3,1)$. Moreover, the polar decomposition $Q = UP$ shows that an arbitrary Lorentz transformation Q can be expressed as the product of a space rotation U by some angle about some axis and a boost P by some velocity along some direction. To make this precise we observe that an arbitrary matrix $U \in SU(2)$ can be expressed as

$$U = a\mathbf{1} + i\boldsymbol{\sigma} \cdot \boldsymbol{b}, \tag{19.48}$$

where the four real parameters a and $\boldsymbol{b}$ are constrained by

$$a^2 + \boldsymbol{b} \cdot \boldsymbol{b} = 1. \tag{19.49}$$

We use (19.36) to get

$$\Lambda(U)_{44} = \tfrac{1}{2}\mathrm{tr}(a1+i\boldsymbol{\sigma}\cdot\boldsymbol{b})(a1-i\boldsymbol{\sigma}\cdot\boldsymbol{b}) = 1, \tag{19.50}$$

$$\Lambda(U)_{4k} = \tfrac{1}{2}\mathrm{tr}(-ia1 + \boldsymbol{\sigma}\cdot\boldsymbol{b})\sigma_k(a1-i\boldsymbol{\sigma}\cdot\boldsymbol{b}) = 0 = \Lambda(U)_{k4}, \tag{19.51}$$

$$\Lambda(U)_{jk} = \tfrac{1}{2}\mathrm{tr}\sigma_j(a1+i\boldsymbol{\sigma}\cdot\boldsymbol{b})\sigma_k(a1-i\boldsymbol{\sigma}\cdot\boldsymbol{b})$$
$$= (a^2 - \boldsymbol{b}\cdot\boldsymbol{b})\delta_{jk} + 2a\varepsilon_{jkl}b_l + 2b_j b_k \tag{19.52}$$

and find by explicit calculation that the non-trivial part of $\Lambda(U)$ is a 3×3 unimodular orthogonal matrix. As such its eigenvalues are 1, $\exp i\Theta$, $\exp -i\Theta$, where Θ is the angle of rotation, so its trace is $1+2\cos\Theta$, while $\Lambda(U)_{jj} = 3a^2 - \boldsymbol{b}\cdot\boldsymbol{b} = 4a^2 - 1$. Thus,

$$a^2 = \cos^2\tfrac{\Theta}{2}, \quad \boldsymbol{b}\cdot\boldsymbol{b} = \sin^2\tfrac{\Theta}{2}. \tag{19.53}$$

Since

$$\Lambda(U)_{jk}b_k = b_j \tag{19.54}$$

we see that $\boldsymbol{b}$ is an eigenvector of $\Lambda(U)_{jk}$ to the eigenvalue 1 and therefore is the axis of rotation, that being the only direction invariant to a rotation. Consequently

$$\boldsymbol{b} = \boldsymbol{n}\sin\tfrac{\Theta}{2}, \quad \boldsymbol{n}\cdot\boldsymbol{n} = 1, \tag{19.55}$$

and finally

$$U = 1\cos\tfrac{\Theta}{2} + i\boldsymbol{\sigma}\cdot\boldsymbol{n}\sin\tfrac{\Theta}{2} = \exp i\tfrac{\Theta}{2}\boldsymbol{\sigma}\cdot\boldsymbol{n} \tag{19.56}$$

describes a rotation by an angle Θ about the direction $\boldsymbol{n}$.

On the other hand, the same procedure applied to P shows that an arbitrary hermitian positive matrix $P \in SL(2,\mathbb{C})$ can be expressed in terms of u and the unit vector $\boldsymbol{m}$ (which are three real parameters) as

$$P = 1\cosh\tfrac{u}{2} + \boldsymbol{\sigma}\cdot\boldsymbol{m}\sinh\tfrac{u}{2} = \exp\tfrac{u}{2}\boldsymbol{\sigma}\cdot\boldsymbol{m}, \tag{19.57}$$

which describes a boost by a velocity $\boldsymbol{v} = v\boldsymbol{m}$, where

$$v = \tanh u. \tag{19.58}$$

In conclusion, we note that the polar decomposition of $SL(2,\mathbb{C})$ is readily generalized as follows: For $SL(n,\mathbb{C})$ we have the polar decomposition (19.45) with $U \in SU(n)$ and the parameter space of P being $\mathbb{R}^{n^2-1}$. Thus, $SL(n,\mathbb{C})$ is simply connected. For $GL(n,\mathbb{C})$ we have (19.45) with $U \in U(n)$

and the parameter space of P (which is no longer unimodular) being $\mathbb{R}^{n^2}$ and the connectivity properties of $GL(n,\mathbb{C})$ are those of $U(n)$ (in particular these groups are *not* simply connected).

For $GL(n,\mathbb{R})$ the matrices in the polar decomposition (19.45) become real:

$$Q=OP, \qquad (19.59)$$

where O is an orthogonal matrix and P is a positive symmetric matrix. Now, the parameter space of P is $\mathbb{R}^{n(n+1)/2}$ and the connectivity properties of $GL(n,\mathbb{R})$ are those of $O(n)$. For $SL(n,\mathbb{R})$ the matrices O and P are in addition unimodular. Then, the parameter space of P is $\mathbb{R}^{(n-1)(n+2)/2}$ and the connectivity properties of $SL(n,\mathbb{R})$ are those of $SO(n)$. In particular for $n>2$ $SL(n,\mathbb{R})$ is doubly connected.

Since $\mathbb{R}$ is unbounded it follows that all of the following groups are non-compact: $GL(n,\mathbb{C})$, $GL(n,\mathbb{R})$, $SL(n,\mathbb{C})$, $SL(n,\mathbb{R})$.

Biographical Sketches

Minkowski, Hermann (1864–1909) was born near Kovno, Russia (now Kaunas, Lithuania). He was educated at the University of Königsberg where he and his friend Hilbert received their doctorates in 1885 under the supervision of Lindemann. He was appointed professor there in 1895 and then at Zürich in 1896, where Einstein was his student. He moved to Göttingen in 1902 and died there. His geometrization of special relativity evoked this comment from Einstein: "Since the mathematicians have invaded the theory of relativity I do not understand it myself anymore."

Klein, Oscar (1894–1977) was born near Stockholm, the son of the chief rabbi of Stockholm. When he was on his way to visit Perrin in France World War I broke out and he was drafted. Starting in 1917 he spent several years at the Bohr Institute in Copenhagen. He received his doctorate at the University College of Stockholm in 1921. He became professor at the University of Michigan in 1923 and returned to Europe in 1925. In 1930 he took over the professorial chair in physics at his alma mater, where he stayed until his retirement in 1962. He was awarded the Max Planck medal in 1959. Here are some concepts named after him in addition to the Klein–Gordon equation: the Klein–Nishina formula, the Klein paradox, the Kaluza–Klein theory.

Gordon, Walter (1893–1939) was born in Apolda, Germany. In 1915 he came to the University of Berlin and received his doctorate there in 1921 from Planck and became the assistant of von Laue in 1922. In 1925 he spent some time with Bragg in Manchester. In 1926 he moved to Hamburg and in 1930 he became a professor at the University of Hamburg. Because of the rise of the Nazis he moved to Stockholm in 1933 and stayed there till his death.

Dirac, Paul Adrien Maurice (1902–1984) was born in Bristol and educated at the universities of Bristol and Cambridge. He did his doctoral work under Fowler, his thesis finished in 1926 and entitled "Quantum Mechanics" was one of the early ones to be submitted anywhere on this subject. In 1928 he discovered the equation named after him that predicted antimatter. In 1930 he published a textbook on quantum mechanics that is used to this day. His appointment to the Lucasian Chair, once held by Newton, came in 1932 when he was just a few months older than Newton. In 1933 he shared the Nobel Prize with Schrödinger. He spent the last years of his life at Florida State University. When asked to summarize his philosophy of physics Dirac wrote "Physical laws should have mathematical beauty."

Proca, Alexandru (1897–1955) was born in Bucharest, Romania. In 1922 he graduated from the Bucharest Polytechnic School as an Electromechanical Engineer. In 1923 he moved to France, in 1925 became a graduate of the Science Faculty, Sorbonne University, Paris. In 1933 he submitted his doctorate thesis to a commission consisting of Perrin, de Broglie, Brillouin and Cotton. In 1943 he moved to Portugal and lectured at the University of Porto.

20

The Poincaré and Liouville groups

In addition to our belief that the laws of Physics are invariant under four-dimensional *rotations* of space-time we believe in invariance under four-dimensional *translations* of space-time. The translations are generated by the four-momentum P_μ consisting of the momentum three-vector $\mathbf{P}=(P_1, P_2, P_3)$ and the energy $P_0=-iP_4$. Since P_μ is canonically conjugate to the position four-vector it can be realized as $-i\partial_\mu$ from which we immediately deduce the commutation relations

$$[P_\mu, P_\nu]=0, \tag{20.1}$$

$$[P_\lambda, M_{\mu\nu}]=i\delta_{\lambda[\nu}P_{\mu]}, \tag{20.2}$$

i.e. P_λ and $M_{\mu\nu}$ form together a 10-dimensional algebra called the **Poincaré algebra**, the four P_λ form an Abelian subalgebra of translations, and transform like *rank one tensors* (four-vectors) under the Lorentz subalgebra formed by the $M_{\mu\nu}$.

We define the **Poincaré group** to be the universal cover of the group of transformations that leave invariant the relative distance between two points (events) in space-time. Thus, the Poincaré group is the semi-direct product of the translation group T_4 and the universal covering group of $SO_0(3,1)$:

$$T_4 \wedge SL(2,\mathbb{C}). \tag{20.3}$$

Definition: Let A_i be the generators of one group and B_j the generators of another group so that, symbolically

$$[A, A]=A, \quad [B, B]=B. \tag{20.4}$$

If

$$[A, B]=0 \tag{20.5}$$

then we say that the group generated by A_i and B_j together is the **direct product** group $\otimes$. But if instead

$$[A, B] = A \tag{20.6}$$

then we say that the A_i and B_j together generate the **semidirect product** group $\wedge$.

Now, what can we say about the Casimirs of the Poincaré group? Since they must commute with all generators they commute with the generators of the Lorentz subgroup, i.e. are Lorentz invariants. Since products of generators of the Poincaré group transform under the Lorentz group as tensors of rank equal to the number of free (not summed over) subscripts, it follows that if *all* subscripts are summed over we have a tensor of rank zero, a scalar, a Lorentz invariant. The only Lorentz invariants that can be formed out of products of the $M_{\mu\nu}$ alone are the two Casimirs of the Lorentz group F_1 and F_2; they fail to commute with P_μ and so fail to be Casimirs of the Poincaré group. The only Lorentz invariant that can be formed out of products of the P_μ alone is $P_\mu P_\mu$; it obviously commutes with P_ρ and so we have our first Casimir of the Poincaré group. It remains to consider structures that contain *both* $M_{\mu\nu}$ and P_ρ of which the simplest is $M_{\mu\nu} P_\rho$. We find that

$$[M_{\mu\nu} P_\rho, P_\alpha] = i\delta_{\alpha[\mu} P_{\nu]} P_\rho \tag{20.7}$$

which vanishes when contracted with $\varepsilon_{\sigma\mu\nu\rho}$. Therefore, the so-called Pauli–Lubanski four-vector

$$W_\alpha \equiv -i\varepsilon_{\alpha\mu\nu\sigma} M_{\mu\nu} P_\sigma /2 \Rightarrow W_0 = \boldsymbol{J}\cdot\boldsymbol{P}, \quad \boldsymbol{W} = \boldsymbol{P}\times\boldsymbol{N} + P_0 \boldsymbol{J} \tag{20.8}$$

commutes with P_μ and we immediately obtain two more candidates for Casimirs of the Poincaré group, namely the Lorentz invariants $W_\mu P_\mu$ and $W_\mu W_\mu$. However, we see from its definition that $W_\mu P_\mu = 0$—this makes it useless as a Casimir but it is important in that it states that W_μ has only *three* independent components. Thus, we find the two Casimirs

$$P^2 = P_\mu P_\mu \quad \text{and} \quad W^2 = W_\mu W_\mu \tag{20.9}$$

the first being a quadratic, the second a quartic, polynomial in the generators. This exhausts the list of Casimir operators as any polynomials of higher degree are not polynomially independent of P^2 and W^2.

To obtain a representation we construct a basis for the states that the Poincaré operators act on. We choose the states to be simultaneous eigenstates of as many commuting operators as possible. We know that the four P_μ commute among themselves and they commute with W_ν. However, the W_ν do not commute among themselves since

$$[W_\alpha, W_\beta] = \varepsilon_{\alpha\beta\mu\nu} P^\mu W^\nu. \tag{20.10}$$

So, we propose to choose our states as simultaneous eigenstates of P^2, W^2, P_1, P_2, P_3, P_0 and some one component of W_μ.

The four objects P_1, P_2, P_3, P_0 are hermitian commuting operators. When simultaneously diagonalized they have real eigenvalues p_1, p_2, p_3, p_0, all ranging continuously over the real line:

$$-\infty \leq p_\mu \leq +\infty, \quad \mu = 1, 2, 3, 0. \tag{20.11}$$

It follows then that P^2 has the eigenvalue $p_\mu p^\mu = \mathbf{p}^2 - p_0^2 = -M^2$, where M^2 is a *real number* that could be positive, negative or zero, corresponding to the four-vector P_μ being **time-like, space-like** or **light-like**. Not surprisingly this leads to three classes of irreducible representations of the Poincaré group.

Class I, P_μ time-like. When P_μ is time-like there exists a special frame of reference in which $\mathbf{p} = 0$, $p_0 = \pm M$, this is the so-called **rest frame**. Note that a time-like, as well as a light-like, four-vector has another Poincaré invariant associated with it, the *sign* of the zeroth component. Thus, when specifying an irreducible representation of the Poincaré group one may need to specify invariants that are not polynomials formed out of the generators.

In this special frame the four-vector W_μ becomes

$$W_0 = 0, \quad \mathbf{W} = \pm M \mathbf{J} \tag{20.12}$$

and we can choose our states to be eigenstates of J_3 (which is proportional to W_3) to the eigenvalue m. Recalling our discussion of $SU(2)$ we have

$$W^2 = \mathbf{W}^2 = M^2 \mathbf{J}^2 \to M^2 j(j+1), \tag{20.13}$$

where, as usual, $2j = 0, 1, 2, 3...$, and $-j \leq m \leq j$. Thus, we find that Class I irreducible representations are specified by M and j—the **rest mass** and **spin** (and the sign of the energy)—this is the familiar description of massive particles and the reason for referring to W_μ as the relativistic spin as it is a Lorentz four-vector equivalent to $\mathbf{J}$ in the rest frame of P_μ.

These representations are unitary representations of the Poincaré group, they are infinite-dimensional possessing $2j+1$ components corresponding to spin j and a continuous infinity of momentum values subject to $p_\mu p_\mu = -M^2$. Most of the particles occurring in nature belong to this class, here are some

spin 0: the π meson, the K meson, the η meson, the Higgs boson (maybe);

spin 1/2: the electron, the muon, the tau meson, the quark, the nucleon;

spin 1: the ρ meson, the ω meson, the W and Z vector bosons;

spin 3/2: Δ, Ω.

Class II, P_μ light-like. When P_μ is light-like there is no rest frame. It is convenient to consider a frame in which $p_1 = p_2 = 0$, $p_3 = p$, $p_0 = \pm p$. Now, two subclasses need to be considered corresponding to $p \neq 0$ or $p = 0$.

Class IIa, $p \neq 0$. Since p is the eigenvalue of P_3 we can take it positive because if it were negative we could make it positive by a rotation by π in the 1–3 plane. The sign of p_0 is, however, again an invariant. Setting for convenience $W_\mu = pK_\mu$ we have in the preferred frame (the $\pm$ signs corresponding to the sign of p_0)

$$K_1 = \pm J_1 - N_2,$$
$$K_2 = \pm J_2 + N_1,$$
$$K_3 = \pm J_3,$$
$$K_0 = J_3,$$
(20.14)

with the commutation relations for the three independent components

$$[K_1, K_2] = 0,$$
$$[K_1, K_0] = -iK_2,$$
$$[K_2, K_0] = iK_1.$$
(20.15)

What is the meaning of these K_1, K_2, K_0? Since $K_0 = \boldsymbol{P} \cdot \boldsymbol{J}/p$ it is the component of angular momentum in the direction of the linear momentum in this frame—it is called **helicity**. K_1 and K_2 on the other hand commute with each other, behave similarly to linear momenta. In fact, if we write the commutation relations thus

$$[K_i, K_j] = 0, \quad [K_j, M_{lk}] = i\, \delta_{j[k} K_{l]},$$
(20.16)

(using $K_0=J_3=M_{12}$) with all indices taking on just the two values 1,2, and compare it with the Poincaré algebra

$$[P_\mu, P_\nu]=0, \quad [P_\mu, M_{\kappa\lambda}]=\mathrm{i}\,\delta_{\mu[\lambda}P_{\kappa]}$$

it is clear that we have an algebra just like the Poincaré algebra but in only two dimensions, that is K_1, K_2 and M_{12} generate the group of isometries in two dimensions

$$E(2)=T_2 \wedge SO(2), \qquad (20.17)$$

where K_1, K_2 generate the two translations, M_{12} the one rotation of the 2-dimensional plane. Since all K_1, K_2, M_{12} are supposed to be hermitian we have the so-called **Euclidean** case as opposed to Minkowski, which accounts for the use of E in the name of the group. To be precise, we will look for representations of $\hat{E}(2)$, the universal covering group of $E(2)$.

So, K_i are components of a two-vector under rotations generated by M_{12}, therefore $K_1{}^2+K_2{}^2$, is an invariant, a Casimir of $\hat{E}(2)$. Its eigenvalue M_K^2 is a real number, a "mass squared" and because of the Euclidean metric it is either positive or zero, and zero only if $K_1=K_2=0$. Therefore, if we denote the eigenvalue of W^2 by w^2 then

$$w^2=p^2 M_K^2 \;\Rightarrow\; 0\leq w^2<\infty, \qquad (20.18)$$

which is why these representations are sometimes called continuous spin representations [compare with (20.13)]. In addition to P_μ and W^2 our states can be eigenstates of $W_0/p=J_3$—the angular momentum in the direction of the linear momentum or helicity—to the eigenvalue λ. To simplify the notation we denote these states by $|\lambda\rangle$ omitting all the other labels identifying the eigenvalues of P_μ and W^2.

Introducing $K_\pm=K_1\pm iK_2$ we have

$$J_3 K_\pm|\lambda\rangle=(\lambda\pm 1)K_\pm|\lambda\rangle \qquad (20.19)$$

so that

$$K_\pm|\lambda\rangle\sim|\lambda\pm 1\rangle,$$
$$\langle\lambda\pm 1|\lambda\pm 1\rangle=\langle\lambda|K_\mp K_\pm|\lambda\rangle=\langle\lambda|K_1^2+K_2^2|\lambda\rangle=M_K^2\langle\lambda|\lambda\rangle. \qquad (20.20)$$

We have two possibilities: M_K^2 is positive so that W_μ is space-like and the norm squared of the above states is positive. Therefore, starting from some integer value of λ we obtain all integer values (positive, negative, or zero), while starting from some half-odd-integer value we obtain all half-odd-integer values (positive or negative). Thus, we have an infinite-dimensional representation of $\hat{E}(2)$, as we should since it is non-compact, and, of course, an infinite-dimensional representation of the Poincaré group.

The other possibility is that $M_K^2=0$ so that W_μ is light-like. In this case we must have

$$K_1=K_2=0. \qquad (20.21)$$

Our states can again be simultaneously eigenstates of J_3 to eigenvalue λ where λ, the helicity, is from the set $0, \pm 1/2, \pm 1,...$ However, we can not shift this value as $K_\pm$ are now identically zero. Thus, we obtain *one-dimensional representations* of $\hat{E}(2)$.

This is a remarkable result: we have found a finite-dimensional unitary representation of a non-compact group. The trick is that we have represented K_1 and K_2, the generators of the translations T_2, the non-compact part of $\hat{E}(2)$, by zero.

If one considers the Poincaré group extended by parity then the representations become two-dimensional as with λ we must include $-\lambda$. This can be seen by noting that the helicity operator $\boldsymbol{J}\cdot\boldsymbol{P}/|\boldsymbol{P}|$ is the scalar product of the angular momentum pseudovector, which is invariant under parity, with a unit vector in the direction of the linear momentum, which is a vector and changes sign under parity.

There are examples of such particles in nature:

$\lambda=\pm 1$: photon, gluon
$\lambda=\pm 2$: graviton (maybe).

Class IIb, p=0. This is a very degenerate case, all components of P_μ have eigenvalue equal to zero. In effect, the Poincaré group becomes the Lorentz group and all the representations are those we found before for the Lorentz group—we know of no such particles in nature.

Class III, P_μ space-like. Again such objects do not occur in nature except in science fiction, they are called *tachyons* and are superluminal, meaning that they move faster than light. We discuss this case for the sake of completeness. Again there is no rest frame but we can consider a preferred frame for which $p_0=p_1=p_2=0$, $p_3=|M|$. In this frame the components of $W_\mu=|M|K_\mu$ are

$$K_1=-N_2, \quad K_2=N_1, \quad K_3=0, \quad K_0=J_3, \qquad (20.22)$$

so that

$$[N_1, N_2] = -iJ_3,$$
$$[N_2, J_3] = iN_1, \qquad (20.23)$$
$$[J_3, N_1] = iN_2,$$

which if it were not for that minus sign in the first line would give $SU(2)$, but actually gives $SL(2,\mathbb{R})$ (see Chapter 2).

Now, the Casimir of this group is

$$W^2/M^2 = K^2 = \mathbf{K}^2 - K_0^2 = N_1^2 + N_2^2 - J_3^2$$
$$= N_+ N_- - J_3(J_3 - 1) \qquad (20.24)$$
$$= N_- N_+ - J_3(J_3 + 1),$$

where

$$N_\pm = N_1 \pm iN_2. \qquad (20.25)$$

As before we denote our states by $|\lambda\rangle$, where λ, the eigenvalue of J_3, can be $0, \pm 1/2, \pm 1, \ldots$, and as before,

$$J_3 N_\pm |\lambda\rangle = (\lambda \pm 1) N_\pm |\lambda\rangle, \quad N_\pm |\lambda\rangle \sim |\lambda \pm 1\rangle, \qquad (20.26)$$
$$\langle \lambda \pm 1 | \lambda \pm 1 \rangle = \langle \lambda | N_\mp N_\pm | \lambda \rangle = [K^2 + \lambda(\lambda \pm 1)] \langle \lambda | \lambda \rangle \qquad (20.27)$$

and we arrive at different possibilities depending on the value of the Casimir invariant K^2.

Positive discrete series: Suppose that for some value of λ, to be called $\lambda_{\min}$, we have

$$N_- |\lambda_{\min}\rangle = 0, \quad \langle \lambda_{\min} | \lambda_{\min} \rangle > 0. \qquad (20.28)$$

Then, $\langle \lambda_{\min} | N_+ N_- | \lambda_{\min} \rangle = [K^2 + \lambda_{\min}(\lambda_{\min} - 1)] \langle \lambda_{\min} | \lambda_{\min} \rangle = 0$ and therefore

$$K^2 = -\lambda_{\min}(\lambda_{\min} - 1). \qquad (20.29)$$

By acting on $|\lambda_{\min}\rangle$ with N_+ we can raise the value of the helicity by one provided the resultant state has nonvanishing norm. We apply N_+ n

times and consider the norm squared of the resultant state:

$$\begin{aligned}
A_n &= \langle \lambda_{\min}|(N_-)^n(N_+)^n|\lambda_{\min}\rangle \\
&= \langle \lambda_{\min}|(N_-)^{n-1}N_-N_+(N_+)^{n-1}|\lambda_{\min}\rangle \\
&= \langle \lambda_{\min}|(N_-)^{n-1}[K^2+J_3(J_3+1)](N_+)^{n-1}|\lambda_{\min}\rangle \\
&= [-\lambda_{\min}(\lambda_{\min}-1)+(\lambda_{\min}+n-1)(\lambda_{\min}+n)]A_{n-1} \\
&= n(n-1+2\lambda_{\min})A_{n-1} \\
&= \frac{n!(n-1+2\lambda_{\min})!}{(2\lambda_{\min}-1)!} \langle \lambda_{\min}|\lambda_{\min}\rangle.
\end{aligned} \qquad (20.30)$$

Thus, A_n is positive as long as $2\lambda_{\min}-1$ is not a negative integer. In this way we obtain the infinite-dimensional representation with states $|\lambda_{\min}+n\rangle$, n a non-negative integer, where $2\lambda_{\min}=1,2,3,\ldots$. This is called the *positive discrete series*.

Negative discrete series. By the same arguments but starting from

$$N_+|\lambda_{\max}\rangle=0, \langle \lambda_{\max}|\lambda_{\max}\rangle>0 \qquad (20.31)$$

we have $K^2=-\lambda_{\max}(\lambda_{\max}+1)$ and we obtain the *negative discrete series* with states $|\lambda_{\max}-n\rangle$, n a non-negative integer and $2\lambda_{\max}=-1,-2,-3,\ldots$.

We also have the trivial one-dimensional representation $N_1=N_2=J_3=0$.

Thus, the quadratic Casimir operator K^2 takes on the *discrete* values $-k(k-2)/4$, $k=1,2,3,\ldots$, where $k=2\lambda_{\min}$ in the positive discrete series and $k=-2\lambda_{\max}$ in the negative discrete series. For values of the quadratic Casimir *not* of the above form no states are annihilated by either N_+ or N_-. In that case starting from some helicity λ we can use N_+ and N_- to obtain $|\lambda\pm n\rangle$, n positive. This results in the integral *continuous* series for which λ is integer and $K^2>0$ (excluding $1/4$) and the half-odd-integral *continuous* series for which λ is half-odd-integer and $K^2>1/4$.

Note that (except for $2\lambda_{\max}=-1$ and $2\lambda_{\min}=1$) $K^2\leq 0$ for the discrete series, i.e. K_μ is light-like or time-like so that the sign of K_0—hence ultimately the sign of λ—is an invariant. For the continuous series $K^2>0$, K_μ is space-like and both signs of λ are allowed.

It is amusing to note that if we let N_1 and N_2 be anti-hermitian then for the hermitian operators $S_1=iN_1$, $S_2=iN_2$, $S_3=J_3$ the commutation relations (19.55) become $[S_i, S_j] = i\,\varepsilon_{ijk}\,S_k$, i.e. the S_i generate $SU(2)$ whose unitary representations are the familiar finite-dimensional representations. In other words, $SL(2,\mathbb{R})$ has non-unitary representations for which $K^2 = N_1^2 + N_2^2 - J_3^2 = -S_1^2 - S_2^2 - S_3^2 = -j(j+1)$, $2j = 0,1,2,\ldots$. This value of K^2 agrees with the value for the positive discrete series and negative discrete series provided

$$j = \lambda_{\min} - 1 = -\lambda_{\max} - 1. \tag{20.32}$$

Thus, for $K^2 = -j(j+1)$ all helicities occur in this sense: helicities from $-\infty$ to $\lambda_{\max} = -j-1$ occur in the negative discrete series specified by $\lambda_{\max}$, these are followed by the helicities from $-j$ to $+j$ corresponding to the finite-dimensional non-unitary representation specified by j and the remaining helicities from $j+1 = \lambda_{\min}$ to $+\infty$ occur in the positive discrete series specified by $\lambda_{\min}$. We show the weight diagram below for $j=2$:

$$\underbrace{\ldots\; -6 \quad -5 \quad -4 \quad -3}_{\text{negative discrete}} \quad \underbrace{-2 \quad -1 \quad 0 \quad 1 \quad 2}_{\text{non-unitary}} \quad \underbrace{3 \quad 4 \quad 5 \quad 6\ldots}_{\text{positive discrete}}$$

It is useful to rephrase our results for the Poincaré group as follows. Because the translation group T_4 is Abelian we can contemplate a space $\mathcal{M}$ whose points consist of $p_\mu = (\boldsymbol{p}, p_0)$, the simultaneous eigenvalues of P_μ. Because the Poincaré group is a semidirect product we can think of the Lorentz operators as acting on that space—thus, e.g., J_k acts by rotating $\boldsymbol{p}$ around axis k, N_k acts by boosting p_μ along the axis k. As a result of all possible actions by the Lorentz generators a point p_μ in the space $\mathcal{M}$ moves to other points and all such points are called the **orbit** of the original point. Some subset of the Lorentz generators might, however, leave a point unchanged, such generators form a group called by mathematicians the **stabilizer**, also the **isotropy** subgroup, while physicists call it the **little group**.

In Class I p_μ is time-like and the constraint $p^2 = \boldsymbol{p}^2 - p_0^2 = -M^2 < 0$ defines in the space $\mathcal{M}$ a two-sheeted hyperboloid, with p_0 positive on one sheet, negative on the other. A point on one sheet stays there under the action of the Lorentz group because the signs of both p^2 and p_0 are Lorentz invariants. Thus, we have two orbits corresponding to the two sheets. Moreover, a representative point on each sheet of the form $\boldsymbol{p}=0$, $p_0 = \pm M$ is obviously left unchanged by the action of the generators $\boldsymbol{J}$, i.e. the little group is $SU(2)$, the universal cover of $SO(3)$.

In Class II p_μ is light-like, the constraint $\mathbf{p}^2-p_0^2=-M^2=0$ defines a two-sheeted cone, with p_0 positive on the "forward light cone", negative on the "backward light cone", and zero at the point where the two cones touch, giving rise to three orbits. Now, the representative point on each sheet of the form $p_1=p_2=0$, $p_0=\pm|p_3|\neq 0$ is obviously left unchanged by J_3 and, not so obviously, by J_1+N_2 and J_2-N_1, so that the little group is $\hat{E}(2)$. For the third orbit, where the two cones touch and we have in addition $p_0=p_3=0$, the little group is obviously the universal cover of all of the Lorentz group.

In Class III p_μ is space-like and the constraint $\mathbf{p}^2-p_0^2=-M^2>0$ defines a one-sheeted hyperboloid, so we have one orbit and the representative point $p_1=p_2=p_0=0$, $p_3\neq 0$ is obviously left unchanged by N_1, N_2, J_3 and the little group is the universal cover of $SL(2,\mathbb{R})$.

The Liouville group

One more word about space-time groups. The Poincaré group is the group composed of transformations that leave invariant the relative distance between two points in space-time. Now, imagine that you change the *scale* that you use for measuring that distance—then if you had two points whose relative distance was zero that distance would remain unchanged under this operation called **dilation**. In general, the relative distance squared between two points would change by some scalar function. The Poincaré group extended by dilations is sometimes called the Weyl group. We denote the generator of dilations by D, it can be realized as $ix_\mu\partial_\mu$ and therefore its commutation relations with Poincaré generators are

$$[D,\ P_\mu]=-iP_\mu, \quad [D,\ M_{\nu\rho}]=0. \tag{20.33}$$

Because the momentum operators fail to be invariant under dilations it follows that the Poincaré Casimir $P_\mu P_\mu$ is not a Weyl Casimir—except for *massless* representations. For the same reason representations for which $W_\mu W_\mu\neq 0$ are unacceptable. The only acceptable representations are those for which both P_μ and W_μ are light-like, corresponding to the discrete helicity one-dimensional irreducible representations (two-dimensional if extended by parity).

Not surprisingly, Maxwell's equations are invariant under this group— in fact they are invariant under four more operations called **special conformal** transformations generated by K_μ that can be realized as

$\mathrm{i}(2x_\mu x_\nu \partial_\nu - x_\nu x_\nu \partial_\mu)$ resulting in the commutation relations

$$[K_\mu, P_\nu] = -2\mathrm{i}(\delta_{\mu\nu}D + M_{\mu\nu}),$$
$$[K_\rho, M_{\mu\nu}] = \mathrm{i}\delta_{\rho[\mu}K_{\nu]},$$
$$[K_\mu, D] = -\mathrm{i}K_\mu, \tag{20.34}$$
$$[K_\mu, K_\nu] = 0.$$

These 15 generators generate the **conformal group**, also called the Liouville group, it is the largest group of transformations that leave invariant Maxwell's equations. Note that 15 is the dimension of the orthogonal group in six dimensions and the unitary group in four dimensions—indeed the conformal group is related to $O(4,2) \cong U(2,2)$.

We have introduced the conformal group in connection with space-time with 3 space and 1 time dimensions. We can consider arbitrary space-time dimensions N by letting the Greek lower case indices range over N values: $\mu = 1, 2, \ldots, N$, and one finds that the conformal group of N-dimensional space-time is isomorphic to $O(N+2)$. To be precise: for m space and n time dimensions the conformal group is

$$O(m+1, n+1), \tag{20.35}$$

except if $m = n = 1$. In two-dimensional space-time the group of conformal transformations is infinite-dimensional being essentially analytic transformations. Thus, in two dimensions x and t consider the variables $z = x + t$ and $z^\# = x - t$. Given a function of z only we can expand it in a Laurent series in z—thus define the generators

$$L_n \equiv -z^{n+1} d/dz, \quad n \in \mathbb{Z}, \tag{20.36}$$

clearly they form the algebra

$$[L_n, L_m] = (m-n)L_{m+n}, \tag{20.37}$$

and another such algebra can be defined for functions of $z^\#$ only.

The subset L_1, L_0, L_{-1} generates a *finite* subgroup isomorphic to $O(2,1)$, which together with the corresponding subset from $z^\#$ combine to form $O(2,1) \otimes O(2,1) \cong O(2,2)$ and this generalizes to higher dimensions.

Next, consider the generators T^a of any of the Lie algebras that we have been discussing:

$$[T^a, T^b] = if^{ab}{}_c T^c, \qquad (20.38)$$

and define

$$T^a{}_m = z_m T^a, \quad m \in \mathbb{Z}. \qquad (20.39)$$

It follows that

$$[T^a{}_m, T^b{}_n] = if^{ab}{}_c T^c{}_{m+n}. \qquad (20.40)$$

The algebra of the L_n is known as the **Virasoro** algebra, the algebra of the $T^a{}_m$ is known as the untwisted affine **Kac–Moody** algebra provided we further generalize these algebras by a **central extension**:

$$[L_m, L_n] = (m-n)L_{m+n} + (c/12)m(m^2-1)\delta_{m+n,0}, \qquad (20.41)$$
$$[T^a{}_m, T^b{}_n] = if^{ab}{}_c T^c{}_{m+n} + km\delta^{ab}\delta_{m+n,0}. \qquad (20.42)$$

Here, a,b range over dim g of the Lie group of the T^a, but m and n range over the integers from $-\infty$ to $+\infty$ so that the number of generators of the Kac–Moody algebra is *infinite*. In unitary representations $T^a{}_n{}^\dagger = T^a{}_{-n}$.

This construction leads from *every* finite-dimensional Lie algebra to the so-called *direct* or *untwisted* affine Kac–Moody algebras—there also exist *twisted* Kac–Moody algebras.

The numbers c and k are called **central elements**, which means that they commute with all the generators. Further, the values of k are quantized:

$$2k/\psi^2 \in \mathbb{Z}_+, \qquad (20.43)$$

where ψ is the long root of g (the non-negative integer $2k/\psi^2$ is called the *level*).

The Virasoro and Kac–Moody together form a semidirect product as

$$[L_m, T^a{}_n] = -nT^a{}_{m+n}, \quad [L_m, k] = 0. \qquad (20.44)$$

Biographical Sketches

Poincaré, Jules Henri (1854–1912) was born in Nancy, France. He studied at the Lycée in Nancy (1862–71), the École Polytechnique (1873–75) and the École de Mines, where he received the degree of ordinary engineer in 1879. At the same time he studied under Hermite and received a doctorate in mathematics from the University of Paris in 1879. Most of the basic ideas of modern topology are due to him. The Poincaré conjecture, formulated at the beginning of the twentieth century, was one of the most important questions in topology. It is the only one of seven Millennium Prize Problems that has been solved (by Perelman in 2002). Poincaré came close to anticipating Einstein's theory of special relativity, showing that the Lorentz transformations form a group, deriving the relativistic addition of velocities and observing that Lorentz transformations could be viewed as rotations in four dimensions with an imaginary time coordinate.

Pauli, Wolfgang Ernst (1900–1958) was born in Vienna, Austria. He studied under Sommerfeld at Munich University where he received his Ph.D. in 1921. He then spent a year at Göttingen as an assistant to Born and a year at Bohr's Institute in Copenhagen. During his years as professor at Hamburg University (1923–28) he formulated the theory of non-relativistic spin introducing his famous 2×2 sigma matrices, and developed the exclusion principle. In 1928 he became a Swiss citizen and a professor at the Zürich Federal Institute of Technology. In 1930 he postulated the neutrino, whose experimental discovery in 1956, two years before his death, evoked this comment "Everything comes to him who knows how to wait". His studies in the 1950s of conservation of quantum properties in interactions paved the way for Lee and Yang to propose parity non-conservation in weak interactions; the experimental discovery of parity non-conservation in 1956 was a deep disappointment to Pauli. He received the Lorentz Medal (1931), the Nobel Prize (1945), the Matteucci Medal (1956) and the Max Planck Medal (1956).

Lubanski, Jòzef Kazimierz (1915–1947) was a Polish physicist. He received his *magister philosophiæ* in 1937 in Wilno, Poland (now Lithuania) and worked for two years as an assistant in theoretical physics at Polish universities. He obtained a grant to travel to Holland and work under Kramers at Leyden. His plans to go to Copenhagen were thwarted by World War II and for some time he worked with Rosenfeld at Utrecht during which time he published a number of papers, mostly in *Physica*, dealing with the properties of mesons. In 1945 he became an assistant at the Laboratory for Aero- and Hydrodynamics of the Technical University at Delft.

Kac, Victor (1943–) was born in Buguruslan, Russia. He received his Ph.D. degree from Moscow State University in 1968. He taught at the Moscow Institute for Electronic Engineering (1968–76). He left the Soviet Union in 1977, became Associate Professor at MIT, was promoted to Full Professor in 1981. He discovered Kac–Moody algebras, found the Kac determinant formula for the Virasoro algebra, discovered an elegant proof of the Macdonald identities. He received the Wigner Medal in 1994.

Moody, Robert Vaughan (1941–) was born in Great Britain. After receiving the Ph.D. degree from University of Toronto in 1966 he joined the Faculty of University of Saskatchewan and was promoted to Full Professor in 1976. In 1989 he moved to the University of Alberta. He discovered Kac–Moody algebras. He received the Wigner Medal in 1994–96 with Kac.

21
The Coulomb problem in n space dimensions

In this problem we start with the two n-vectors $\boldsymbol{R}$ and $\boldsymbol{P}$ that form the **Heisenberg algebra**

$$[R_j,\, R_k]=0, \quad [P_j,\, P_k]=0, \quad [R_j,\, P_k]=\mathrm{i}\delta_{jk}, \tag{21.1}$$

with lower case Latin subscripts running from 1 to n. The non-relativistic Hamiltonian (for a particle of mass m and charge $-e$ at the point $\boldsymbol{R}$ moving in the Coulomb potential due to a charge e at the origin) is

$$H = \boldsymbol{P}\cdot\boldsymbol{P}(2m)^{-1} - e^2 R^{-1}, \tag{21.2}$$

where R is the magnitude of $\boldsymbol{R}$,

$$R = (\boldsymbol{R}\cdot\boldsymbol{R})^{1/2}. \tag{21.3}$$

The orbital angular momentum

$$L_{jk} = R_{[j}P_{k]} = -L_{kj} \tag{21.4}$$

has $n(n-1)/2$ independent components and generates $so(n)$, the orthogonal algebra in n dimensions. As a consequence of (21.1) we have the following commutation relations

$$[L_{jk},\, R_m] = \mathrm{i}\delta_{m[j}R_{k]}, \tag{21.5}$$

$$[L_{jk},\, p_m] = \mathrm{i}\delta_{m[j}R_{k]}, \tag{21.6}$$

$$[L_{jk},\, L_{mn}] = \mathrm{i}(\delta_{m[j}L_{k]n} - \delta_{n[j}L_{k]m}), \tag{21.7}$$

that is $\boldsymbol{R}$ and $\boldsymbol{P}$ are tensors of rank one (vectors), while L_{jk} is an antisymmetric tensor of rank two under the transformations generated by L_{jk} (rotations in n dimensions).

Obviously H is a tensor of rank zero, a scalar:

$$[L_{jk}, H]=0, \qquad (21.8)$$

and therefore the L_{jk} are constants of the motion since H generates the evolution of the system in time.

Next, we introduce the Lenz–Runge vector

$$\boldsymbol{A}=\boldsymbol{R}R^{-1}+\boldsymbol{B}, \qquad (21.9)$$

where

$$\begin{aligned}B_j&=\frac{1}{2me^2}(L_{kj}P_k+P_kL_{kj})\\&=\frac{1}{2me^2}[2L_{kj}P_k+\mathrm{i}(1-n)P_j] \qquad (21.10)\\&=\frac{1}{2me^2}[2P_kL_{kj}-\mathrm{i}(1-n)P_j].\end{aligned}$$

We are interested in the square of the Lenz–Runge vector

$$\boldsymbol{A}\cdot\boldsymbol{A}=1+R^{-1}\boldsymbol{R}\cdot\boldsymbol{B}+\boldsymbol{B}\cdot\boldsymbol{R}R^{-1}+\boldsymbol{B}\cdot\boldsymbol{B}. \qquad (21.11)$$

Now,

$$\begin{aligned}\boldsymbol{B}\cdot\boldsymbol{B}&=\frac{1}{4m^2e^4}[2L_{kj}P_k+\mathrm{i}(1-n)P_j][2P_mL_{mj}-\mathrm{i}(1-n)P_j]\\&=\frac{1}{4m^2e^4}[4L_{kj}P_kP_mL_{mj}+(1-n)^2\boldsymbol{P}\cdot\boldsymbol{P}]\\&=(\boldsymbol{P}\cdot\boldsymbol{P}/m^2e^4)[C_2(n)+(1-n)^2/4],\end{aligned}$$

where we have used

$$\begin{aligned}L_{kj}P_kP_mL_{mj}&=R_{[k}P_{j]}P_kP_mL_{mj}=\boldsymbol{R}\cdot\boldsymbol{P}P_jP_mL_{mj}-R_jP_m\boldsymbol{P}\cdot\boldsymbol{P}L_{mj}\\&=L_{mj}\boldsymbol{P}\cdot\boldsymbol{P}L_{mj}/2=\boldsymbol{P}\cdot\boldsymbol{P}C_2(n),\end{aligned}$$

where

$$C_2(n)=L_{mj}L_{mj}/2 \qquad (21.12)$$

is the quadratic Casimir operator of $so(n)$. Next

$$R^{-1}\boldsymbol{R}\cdot\boldsymbol{B}+\boldsymbol{B}\cdot\boldsymbol{R}R^{-1}$$

$$=\frac{1}{2me^2}[2R^{-1}R_jP_kL_{kj}-\mathrm{i}(1-n)(R^{-1}\boldsymbol{R}\cdot\boldsymbol{P}-\boldsymbol{P}\cdot\boldsymbol{R}R^{-1})+2L_{kj}P_kR_jR^{-1}]$$

$$=-\frac{1}{2me^2}[2R^{-1}L_{kj}L_{kj}+\mathrm{i}(1-n)(R^{-1}\boldsymbol{R}\cdot\boldsymbol{P}-\boldsymbol{P}\cdot\boldsymbol{R}R^{-1})]$$

$$=-\frac{2}{me^2R}[C_2(n)+(1-n)^2/4],$$

where we have used

$$R^{-1}\boldsymbol{R}\cdot\boldsymbol{P}-\boldsymbol{P}\cdot\boldsymbol{R}R^{-1}=[R^{-1}R_j,P_j]=\mathrm{i}\tfrac{\partial}{\partial R_j}(R_jR^{-1})$$

$$=\mathrm{i}(R^{-1}\tfrac{\partial}{\partial R_j}R_j+R_j\tfrac{\partial}{\partial R_j}R^{-1})=\mathrm{i}R^{-1}(n-1).$$

Thus,

$$\boldsymbol{A}\cdot\boldsymbol{A}=1+\frac{2H}{me^4}[C_2(n)+(1-n)^2/4], \qquad (21.13)$$

which shows that $\boldsymbol{A}\cdot\boldsymbol{A}$ is a constant of the motion (actually $\boldsymbol{A}$ itself is). If we rescale $\boldsymbol{A}$ by introducing

$$\boldsymbol{D}\equiv\sqrt{\frac{me^4}{-2H}}\boldsymbol{A} \qquad (21.14)$$

then $\boldsymbol{D}$ is hermitian on the subspace of negative values of H and we have

$$-H=me^4[2C_2(n)+2\boldsymbol{D}\cdot\boldsymbol{D}+(1-n)^2/2]^{-1}. \qquad (21.15)$$

Now comes a remarkable fact involving the orthogonal algebra in $n+1$ dimensions, namely the observation that

$$[D_k,\ D_j]=\mathrm{i}L_{kj}. \qquad (21.16)$$

The proof of (21.16) is completely straightforward involving the same type of manipulations that were used to calculate $\boldsymbol{A}\cdot\boldsymbol{A}$. We also have

$$[L_{jk},\ D_m]=\mathrm{i}\delta_{m[j}D_{k]}, \qquad (21.17)$$

which simply states that $\boldsymbol{D}$ is an n-vector.

Hence, if we define
$$L_{k,n+1}=-L_{n+1,k}=D_k \qquad (21.18)$$
then we can combine all the commutators involving L_{kj} and D_m into
$$[L_{\alpha\beta}, L_{\mu\nu}]=i(\delta_{\mu[\alpha}L_{\beta]\nu}-\delta_{\nu[\alpha}L_{\beta]\mu}), \qquad (21.19)$$
where lower case Greek subscripts range from 1 to $n+1$, i.e. the L_{jk} and D_m together generate $so(n+1)$. Moreover,
$$\begin{aligned} C_2(n)+\boldsymbol{D}\cdot\boldsymbol{D} &= (L_{kj}L_{kj}+L_{k,n+1}L_{k,n+1}+L_{n+1,k}L_{n+1,k})/2 \\ &= L_{\alpha\beta}L_{\alpha\beta}/2 = C_2(n+1) \end{aligned} \qquad (21.20)$$
so that (21.15) can be rewritten as
$$-H=me^4[2C_2(n+1)+(1-n)^2/2]^{-1}. \qquad (21.21)$$

To determine the values of the quadratic Casimir we note that we are dealing with the *orbital* realization of the $so(n+1)$ algebra in which the generators are realized as
$$L_{\alpha\beta}=-ix_{[\alpha}\partial_{\beta]}. \qquad (21.22)$$
In this realization the quadratic Casimir becomes
$$\begin{aligned} C_2(n+1) &= -(x_\alpha\partial_\beta x_\alpha\partial_\beta - x_\alpha\partial_\beta x_\beta\partial_\alpha) \\ &= -x_\alpha(x_\alpha\partial_\beta+\delta_{\alpha\beta})\partial_\beta + x_\alpha\partial_\beta(\partial_\alpha x_\beta - \delta_{\alpha\beta}) \\ &= -x_\alpha x_\alpha\partial_\beta\partial_\beta + x_\alpha\partial_\alpha\partial_\beta x_\beta - 2x_\alpha\partial_\alpha \\ &= -x_\alpha x_\alpha\partial_\beta\partial_\beta + x_\alpha\partial_\alpha(x_\beta\partial_\beta+\delta_{\beta\beta}) - 2x_\alpha\partial_\alpha \\ &= -x_\alpha x_\alpha\Delta + N(N+n-1), \end{aligned} \qquad (21.23)$$
where
$$\Delta \equiv \partial_\beta\partial_\beta \qquad (21.24)$$
is the Laplacian in $n+1$ dimensions and
$$N \equiv x_\alpha\partial_\alpha \qquad (21.25)$$
gives the degree of homogeneity when acting on homogeneous functions of the x_α. As a consequence, we have that harmonic functions on $\mathbb{R}^{n+1}$ of

degree of homogeneity l_{n+1} are eigenfunctions of the quadratic Casimir to the eigenvalue

$$l_{n+1}(l_{n+1}+n-1). \tag{21.26}$$

It follows that the expression for the eigenvalue E of the Hamiltonian becomes

$$-E=2me^4(2l_{n+1}+n-1)^{-2}, \tag{21.27}$$

where

$$l_{n+1}=0,1,2,...\text{for } n\geq 2, \tag{21.28}$$

$$l_2=0,\pm 1,\pm 2,....$$

In particular, for $n=3$ we have

$$-E=\frac{me^4}{2(l_4+1)^2} \tag{21.29}$$

and we recognize the non-relativistic formula for the energy levels of the hydrogen atom with l_4+1 the principal quantum number.

Thus, the energy levels depend only on the integer l_{n+1} and the degree of degeneracy of each level is given by the degree of degeneracy of the quadratic Casimir of $so(n+1)$ in the orbital realization, which is

$$\frac{2(l_{n+1}+n-1)!}{(n-1)!l_{n+1}!}-\frac{(l_{n+1}+n-2)!}{(n-2)!l_{n+1}!}. \tag{21.30}$$

The ground state corresponds to $l_{n+1}=0$ and we see that it is non-degenerate. Equation (21.30) gives for the first few values of n the following

$$n=1 \rightarrow 2-\delta_{l_2,0}$$

$$n=2 \rightarrow 2l_3+1$$

$$n=3 \rightarrow (l_4+1)^2$$

and we recognize the dimensions of the corresponding irreducible representations of the rotation group except for $n=1$, where all irreducible representations are one-dimensional since $so(2)$ is Abelian.

The case $n=1$ has some amusing properties. We have

$$-E = \frac{me^4}{2}(l_2)^{-2}, \qquad (21.31)$$

i.e. the non-degenerate ground state lies infinitely deep, while all the other states are of finite energy and two-fold degeneracy. Here, the Lenz–Runge vector is $x/|x|$ and it is the generator of the $so(2)$. The two-fold degeneracy of all but the ground state is explained by the fact that the Hamiltonian commutes both with the Lenz–Runge vector $x/|x|$ and the parity operator π, but $x/|x|$ and π anticommute.

In conclusion we comment, in connection with the name of the Lenz–Runge vector, on the strange ways of history. The Lenz–Runge vector was first shown to exist by the Swiss mathematician Jacob Hermann (1678–1733) in the eighteenth century. It was rediscovered by Johann Bernoulli, by Laplace, by Hamilton and by Gibbs. Runge used Gibbs derivation as an example in a popular German textbook on vectors, which was referenced by Lenz in his paper on the hydrogen atom.

Biographical Sketches

Coulomb, Charles Augustine de (1736–1806) was born in Angoulême, France. He studied in Paris at Collège des Quatre-Nations. After a long period of service in Martinique he returned to France in 1779. In 1795 he was elected a member of the new Institut de France. Best known for the torsion balance for measuring the force of magnetic and electric attraction. The SI unit of electric charge is named after him.

Heisenberg, Werner Karl (1901–1976) was born in Würzburg, Germany. He studied at the University of Munich under Sommerfeld and Wien and the University of Göttingen under Born, Franck and Hilbert before becoming Professor of Physics at Leipzig University (1927–41). From 1941 to 1945 he was professor at Berlin University and Director of the Kaiser Wilhelm Institute. After the end of WWII that institute was renamed the Max Planck Institute and Heisenberg was its director. In 1958 he became professor at the University of Munich. He received the 1932 Nobel Prize and the 1933 Max Planck Medal. He is best known for the uncertainty principle. During his tenure at Leipzig the list of scholars there included Felix Bloch, Ugo Fano, Siegfried Flügge, Friedrich Hund, Robert Mulliken, Rudolf Peierls, George Placzek, Isidor Isaac Rabi, Fritz

Sauter, John Slater, Edward Teller, John van Vleck, Victor Weisskopf, Carl von Weizsäcker and Gregor Wentzel.

Lenz, Wilhelm (1888–1957) was born in Hamburg, Germany. He studied at the University of Munich under Sommerfeld and received his doctorate in 1911. During WWI he served as a radio operator in France and received the Iron Cross in 1916. From 1921 until his retirement in 1956 he was Professor of Theoretical Physics and Director of the Institute for Theoretical Physics at the University of Hamburg. At Hamburg he trained Ernst Ising and Hans Jensen, and his assistants there included, Pauli, Jordan and Unsold. He is best known for his invention of the Ising model.

Runge, Carl David Tolmé (1856–1927) was born in Bremen, Germany. He studied at the University of Berlin under Weierstrass and received his Ph.D. in 1880. He was professor at University of Hannover (1886–1904) and University of Göttingen (1904–25). He was doctoral advisor to Max Born. He is best known for the Runge–Kutta method.

Bibliography and References

Barut, A. O. and R. Raczka (1986, reprinted 2000) *Theory of Group Representations and Applications* (World Scientific, Singapore).
Biedenharn, L. C. and H. Van Dam, eds. (1965) *Quantum Theory of Angular Momentum* (Academic Press, New York).
Cahn, R. N. (1984) *Semi-Simple Lie Algebras and Their Representations* (Benjamin/Cummings, Menlo Park).
Cartan, É. (1938) *Lecons sur la Théorie des Spineurs* (Hermann, Paris).
Dyson, F. J. (1966) *Symmetry Groups in Nuclear and Particle Physics* (W. A. Benjamin, New York).
Edmonds, A. R. (1957) *Angular Momentum in Quantum Mechanics* (Princeton University Press, Princeton).
Gel'fand, I. M., R. A. Minlos, and Z. Y. Shapiro (1963) *Representations of the Rotation and Lorentz Groups and their Applications* (Pergamon Press, New York).
Gilmore, R. (1974) *Lie Groups, Lie Algebras and Some of Their Applications* (John Wiley & Sons, New York).
Gourdin, M. (1982) *Basics of Lie Groups* (Editions Frontieres, Gil sur Yvette, France).
Hammermesh, M. (1962) *Group Theory and its Applications to Physical Problems* (Addison-Wesley, Reading).
Onishchik, A. L. (Ed.) (1988) *Lie Groups and Lie Algebras I*, Encyclopedia of Mathematical Sciences, Vol. 20 (Springer-Verlag).
Sternberg, S. (1994) *Group Theory and Physics* (Cambridge University Press, Cambridge).
Tung, Wu-Ki (1985) *Group Theory in Physics* (World Scientific).
Weyl, H. (1939) *The Classical Groups, their Invariants and Representations* (Princeton University Press, Princeton, NJ).
Wigner, E. P. (1959) *Group Theory and its Application to Quantum Mechanics of Atomic Spectra* (Academic Press, New York).
Wybourne, B. G. (1974) *Classical Groups for Physicists* (Wiley, New York).
Bargmann, V. *Irreducible Unitary Representations of the Lorentz Group*, Ann. Math. **48**, No. 3, 568–640 (July 1947).

Goddard, P. and D. Olive, *Kac-Moody and Virasoro Algebras in Relation to Quantum Physics*, Int. J. Mod. Phys. **A1** 303–414 (1986).

Pauli, W. *Über das Wasserstoffspectrum vom Standpunkt der neuen Quantummechanik*, Z. Phys. **36**, 336–363 (1926).

Index

Abel, 7
Abelian, 1–4, 11–12, 19, 21, 24, 49, 129
adjoint representation, 11, 22, 26, 61, 86, 121–2, 128, 131, 138, 140, 160
algebra
  classical, 148
  Clifford, 47
  composition, 74–5, 82
  division, 74–5
  exceptional, 143, 148–9
  Heisenberg, 189
  Kac–Moody, 186
  Lie, 10–12, 22, 31, 102–3, 129, 186
  orthogonal, 42, 48, 89
  Poincaré, 175, 179
  Virasoro, 186
ambivalence, 104, 160
angular momentum, 9, 22, 28, 32, 58, 87, 125, 162, 164
anomaly, 106
automorphism, 71–3, 86–7, 104, 121, 160–1
  inner, 71–2
  outer, 71–3

Cartan, 17, 43, 129, 142–3, 148
  basis, 129
  criterion, 11
  integers, 151
  matrix, 17, 151
  metric tensor, 11, 13, 15 17, 31, 43, 129, 131, 133, 137
  subalgebra, 90, 129, 133, 137, 140
Casimir, 39
  operator, 17, 30–1, 89–92, 104–6, 165–7, 176, 179, 181–2, 184, 191–3
Cayley, 74, 83
Cayley–Hamilton theorem, 83, 90, 105

center, 24, 56, 68–70, 72, 103
Clebsch, 40
Clebsch–Gordan coefficient, 35–6, 39, 166
Clifford, 44, 52
  numbers, 43–7, 55, 58, 74, 76–7
compact, 13–19, 31, 43, 96, 102
connected, 20
  simply, 24, 98–9, 165, 170–3
Coulomb, 194
covering, 25, 31–2, 43, 57, 165, 170–1, 175, 179, 184
cycle, 109–11

de Sitter, 94, 100
Dirac, 43, 168, 174
dual, 64
  anti-self, 64–7
  self, 64–7
Dynkin, 73, 151
  diagram, 71, 151–61

Euler, 17, 82
  angles, 9, 163

faithful, 70
folding, 160–1
Frobenius, 83
  theorem, 74, 80
fundamental representation, 136–8, 142–3

Galois, 1, 7
Gordan, 40
Gordon, 168, 174
Graves, 74, 82, 83
group, 1
  classical, 93, 96
  conformal, 186
  cyclic, 2

group (*cont.*)
  general linear, 93, 102–3
  generalized orthogonal, 94
  generalized unitary, 96
  Lie, 9, 22, 31, 165
  Liouville, 94, 184–5
  little, 183–4
  Lorentz, 94, 162–4, 167–8, 180, 183–4
  orthogonal, 20, 42, 89, 93, 97, 98, 163, 185
  Poincaré, 175–8, 180, 181, 183–4
  symmetric, 107–16
  symplectic, 93, 95
  unitary, 93, 96, 102, 108, 185
  unitary symplectic, 96–7
  Weyl, 134, 145, 157, 184

Hamilton, 74, 82
Heisenberg, 194
helicity, 178–84
homomorphism, 2, 23–4, 99, 170
Hurwitz, 82
  theorem, 74

integrity basis, 91, 106
isomorphism, 2, 23, 47, 49–50, 67, 69–71, 74, 78, 83, 99, 103, 104, 119, 149, 159, 167, 170, 185
  of $Spin(n)$ groups, 73

Jacobi, 7
  identity, 7, 130

Kac, 188
Klein, 168, 173
Kronecker, 119, 123, 125–8

Lenz, 195
Lenz–Runge vector, 190, 194
Lie, 17
Liouville, 94, 100
Lorentz, 94, 100, 175–7, 183
  transformations, 9, 163, 170, 171
Lubanski, 187

Maxwell, 94, 100, 169, 184
Minkowski, 163–4, 173, 179
Moody, 188
multiplicity, 39, 87
  inner, 39
  outer, 39

non-compact, 15–17, 31, 94–6, 163–7, 173, 180

octonions, 74, 79–81, 86–7
orthogonal matrix, 19, 20, 42

partition, 87, 115, 123–4
Pauli, 87, 188
Pauli–Lubanski vector, 176
permutation, 107, 109–10
Pfaff, 92
Poincaré, 176, 186–7
polar decomposition, 171–3
Proca, 168, 174

quaternion, 3, 74, 78, 81, 87, 96–7

Racah, 87–8
rank, 90, 129
reduced matrix element, 39, 166
reflections, 19–20, 72, 90, 157, 160
roots, 130–5, 142–9, 157
  simple, 135–7, 140–50, 151, 156–7, 159–60
rotations, 18–23, 31, 58, 68, 72, 91, 162–4, 171–2
Runge, 195

Schur, 52
  lemma, 48, 54, 92
semispinor, 49–51, 53–7, 61–7, 72, 73, 92, 160
shift operators, 27
simply laced, 157, 161
spherical tensor, 37–8
spinor, 31, 43–5, 47–8, 53–7, 58–61, 65–6, 143

Standard Model, 106, 157
symplectic, 54, 57, 66, 70, 104

tachyon, 180
Thomas, 101
   precession, 95, 164

unitary matrix, 8, 102, 108
unitary representation, 29, 30, 31, 42, 44, 165, 168, 178, 180, 183

weight, 27, 131–5, 139, 142, 183
   highest, 27, 134–8
   fundamental, 136, 142–7, 149
Weyl, 40, 168, 184
Wigner, 40, 104
Wigner–Eckart theorem, 39, 166

Young, 112, 116
   pattern, 112–27
   tableaux, 112–15